GEOGRAPHIC PERSPECTIVES
AFGHANISTAN

GEOGRAPHIC PERSPECTIVES

AFGHANISTAN

Eugene J. Palka

UNITED STATES MILITARY ACADEMY AT WEST POINT

The **McGraw·Hill** Companies

Book Team

Vice-President & Publisher   *Jeffrey L. Hahn*
Managing Editor   *Theodore O. Knight*
Director of Production   *Brenda S. Filley*
Developmental Editor   *Ava Suntoke*
Designer   *Charlie Vitelli*
Graphics   *Michael Campbell*
Typesetting Supervisor   *Juliana Arbo*
Proofreader   *Julie Marsh*
Cartography   *Carto-Graphics*

McGraw-Hill/Dushkin

A Division of The McGraw-Hill Companies

Cover: Copyright © 2004 Getty Images
Cover Design *Nancy Norton*

Palka, Eugene J.
Geographic perspectives: Afghanistan/ Eugene J. Palka
New York: McGraw-Hill/Dushkin, © 2004.
p. : cm.
I. Afghanistan—Geography. 1. Series.
915.81
0-07-294009-3

Copyright ©2004 by McGraw-Hill/Dushkin, a division of the McGraw-Hill
Companies, Inc., Guilford, Connecticut 06437
ISBN: 0-07-294009-3   ISSN: 1544-8029

Printed in the United States of America

Contents

For the military personnel and members of nongovernmental organizations who have served in Afghanistan and have made a concerted effort to bring peace, stability, and prospects of a better life to the people of that country

Preface

September 11, 2001, is a day that will be etched in the memory of this generation of Americans for all time, and may well prove to be a turning point in the history of our nation. The direction we follow will depend on our national resolve. The aftermath of the terror created has left America anxious to act, yet looking for answers. Initially, we asked who could have done this horrible act of terror? Why did they do it? And, perhaps most importantly, how do we stop these acts in the future?

Now, two years after this national tragedy, we are at war on two fronts in our campaign against terrorism. We still have more questions than answers, but one person and one place have received much of the focus of our national attention—Osama bin Laden and Afghanistan. Bin Laden is a rich, mysterious individual of Saudi Arabian descent who has a sworn hatred for America couched within his dogmatic version of an ultra-fundamentalist Islamic religious belief. The world was introduced to bin Laden as he became a notable figure while fighting the former Soviet Union in Afghanistan. After more than a year of military action, which has freed Afghanistan from the grip of the Taliban, his whereabouts are unknown and he continues to elude all our efforts to bring him to justice. Although we have restored a measure of security to a war-torn and impoverished country, we still have much to do in order to secure the peace and assist a new government in attaining a level of stability necessary to effectively lead the country into the future.

Afghanistan and its people are not well known or understood by Americans. How should we deal with this abjectly poor country nearly half way around the world? It is a rugged land with 28 million people, of which millions had fled, seeking refuge in adjoining countries to escape the harsh rule of the Taliban. Again, more questions arise. Who were the Taliban? What did they represent? And, why were they our enemies?

As the level of conflict decreases, the coalition forces and the new government of Afghanistan strive to build a stable, self-governing nation from the rubble left from years of harsh and repressive rule by the Taliban, which came on the heels of a civil war, preceded by a decade of war with the former Soviet Union. The issues hindering current nation-building efforts are even deeper than the political challenges of the past, even though some of those issues continue to stifle progress. To begin, one must understand the landscape and the resources it provides in support of the population. As you will read, the available resources in water, energy, arable land, and wealth-generating minerals

Source: John Wiegand

A couple of boys enjoy the view from the top of a communal dwelling in the Panjsher Valley. Villages are typically constructed of sun-baked mud in this part of Afghanistan.

are critically limited, when compared to demands of the population. So, the first important purpose for this book is to describe the physical environment of Afghanistan because many of the causes of instability in the country are rooted in the land's foreboding physical environment.

The natural environment becomes the foundation on which the human footprint can be inscribed, and it is from here we can better understand how the people will influence reestablishing peace and stability in Afghanistan. For people from Western cultures this may be the most difficult part of the puzzle. First, our cultures and perspectives are very different. We can study the people, their long established tribal organizations, their religious and cultural values, and the impacts of years of Taliban oppression. But it is difficult for us to see these through the eyes of the Afghan people. Consider the concepts of peace and stability inherent in the American political values and codified in our constitutional system. These values may be best seen in the Declaration of Independence when in defining the rights of people it states,

> … that they are endowed by their Creator with certain unalienable Rights,
> among these are, Life, Liberty, and the pursuit of Happiness.

I believe these values hold true today, and as President Bush has expressed in speeches describing the goals for a new Afghanistan, these are

Preface

Source: Eugene Palka

A young boy from the village of Charikar poses with his
family bicycle.

values common to all people. Americans are blessed to live in a resource-rich
land where life and liberty are well established and, honestly, people can go
beyond in their pursuit of happiness. For the typical Afghan family, however,
securing clean water, acquiring sufficient food for the day, being safe from
disease, and having protection from attack are daily challenges in their lives.
The people seek a government that can feed them and provide the very basics
for life. Only after the fundamental life-sustaining problems have been solved
can a stable government and a secure peace be established. In a sense, the
Taliban were symptoms of deeper problems in the environmental security of
Afghanistan and we have yet to solve those problems. Afghanistan's history
proves that people who are hungry, sick, and whose children routinely die
before they are two years old, will accept even repression with the promise of
a better life. As you read about the people of Afghanistan, and if you try to consider the perspective of a people who mostly live in abject poverty, the real
magnitude of securing a stable and secure government will come into better
focus.

The authors of this book are uniquely qualified to offer a special perspective on Afghanistan, the land, its people, and securing a lasting peace for the country. First, they are trained geographers, academically and experientially qualified to examine the country over the gamut of physical and cultural subfields. But more, they also are experienced military officers, who are well traveled and in some cases, even have experience on the ground in Afghanistan. As military geographers, some can also add focus to unique military and strategic concerns with the geography of this particular place.

Our goal is to offer a complete, but not exhaustive source of information about Afghanistan and its people. In considering this project, a review of the existing literature was conducted and we found much data, some good, some not very accurate. What was missing was a synthesis of data that could lead to a conceptual understanding of the critical issues related to our security concerns with Afghanistan. This publication does not propose a strategy or policy. The latter must necessarily follow the secession of hostilities and is the focus of others. This publication does provide a view of a place and its people, with perspectives that consider the culture, history, and physical environment. It is intended as a guide and reference for anyone wanting to know more about Afghanistan, a fascinating and diverse country within Central Asia.

—Wendell C. King

1

Introduction

Eugene J. Palka

Key Points

- Afghanistan has been a focus of U.S. and international interest since 9-11.
- The country exhibits tremendous physical and cultural diversity within its borders.
- Regional geography reveals the underlying processes that contribute to Afghanistan's current physical and cultural patterns.

G eography is one of the broadest academic disciplines that has maintained a constant presence within university curricula in America. The breadth of the discipline is a feature that enables practitioners to examine a wide range of problems, at various scales, at different points in time, and across disciplines—from a unique spatial perspective.

Figure 1.1 depicts this symbiotic relationship between geography and other academic disciplines. Where geography overlaps with other disciplines, distinct geographic subfields emerge. While each of these systematic geographies can be studied individually, they are also routinely examined in a collective fashion within the context of particular places or regions. Thus, by definition, most regional geography is both interdisciplinary and multidisciplinary. Geography is not, however, defined by the overlaps with other disciplines, but by its unique spatial perspective, methodologies, tools, and techniques.

Regional geography is one of the discipline's most important overarching methods. Hart (1982) went so far as to claim that regional geography is the "highest form of the geographer's art." As an approach for geographic studies, the regional method is best described as a synthesis of all of the pertinent subfields of the discipline applied to a specific region. All regions have area, location, and boundaries and are based on whatever criteria geographers choose to define them. As organizing principles, regions refer to areal extents of the earth's surface that enable the geographer to compare and contrast

Figure 1.1 The Relationship Between Regional and Systematic Geography

Source: Adapted from De Blij, Muller, and Palka, 2003

different places so that one can ultimately reach a degree of proficiency in areal differentiation. Simply put, the latter enables one to distinguish one place from all others within a global context so that we can better understand the physical and human world in which we live.

The intent herein is to provide a regional geography of Afghanistan, a state that has been in the American consciousness for the past year and a half. Shortly after the September 11, 2001, terrorist attacks on the United States, the U.S. military embarked on Operation Enduring Freedom, with the initial intentions of destroying Osama bin Laden's network of terrorist training camps, disassembling al-Qaeda's infrastructure, and unseating the Taliban government. The operation was oriented on Afghanistan, a mysterious, war-torn country located more than 11,000 kilometers (6,875 miles) away from New York's twin towers. The long-term goals were to bring peace and stability to Afghanistan and Central Asia, with hopes of freeing a suppressed people, improving human rights, promoting economic development, and fostering political cooperation.

Source: John Wiegand
A local contract worker at Bagram airfield.

Prior to the tragic events of 9-11, America's interest in Afghanistan was perhaps more related to the Soviet involvement within the country from 1979 to 1989, rather than with the country itself. Much of the general public's current knowledge of Afghanistan has stemmed from television and media coverage of military operations that have been ongoing for more than a year. Despite popular perceptions about the inhospitable nature of the country, cultures have been rooted in the space of present-day Afghanistan for several thousand years (Palka 2001).

Afghanistan provides an unusual case where military operations were conducted across an entire spectrum from peace and humanitarian missions to combat and within a context where both the physical and human geography have posed fundamental challenges to Coalition forces and nongovernmental agencies. The country's landlocked location in the interior of the continent and contiguity to the Himalaya Mountains contribute to climatic diversity and an assortment of severe meteorological events (such as droughts, or ice, snow, and dust storms), while its location relative to plate boundaries results in intense seismic activity and rugged terrain. Meanwhile, the country's history of conflict, ethnolinguistic diversity, political instability, poverty, and permeable borders present a complex cultural scene.

Source: William A. Jones

Citizens young and old gather and assist in the Herat province canal de-silting project. The canals were neglected over the past seven years. This Coalition Joint Civil Military Operations Task Force (CJCMOTF) project will allow the people of Herat to irrigate their crops during the upcoming growing season and for seasons to come.

In order to provide academics, students, civil and military planners and leaders, and government officials with current, accurate, and relevant information on the distinguishing characteristics of Afghanistan, we have integrated material from various subfields of geography and synthesized the most important features of each to create a concise and balanced picture. As depicted in the subfields in Figure 1.1, this regional geography considers Afghanistan's geomorphology, climatology, and biogeography; its cultural, historical, political, economic, urban geography; and the population and medical subfields to be most pertinent for this work. We begin with a focus on the natural environment, describing various aspects of physical geography. Next we address the inhabitants and their way of life using a human geographic framework. Ultimately, we focus on the interaction between people and their natural surroundings and identify the unique cultures that have emerged within the country.

We did not intend to create a definitive geography of a physically and culturally diverse place that is roughly the size of Texas. Rather, our hope is to provide a geographic portrait that can serve as a point of departure for more detailed studies and to offer a well-organized, informative, easy-to-use reference that fills an existing void for people who have a vested interest in or curiosity about Afghanistan.

Afghanistan: A Geographic Portrait

Source: Preston Cheeks

A boy from the village Mirogul watches as soldiers provide aid during a Medical Civilian Assistance Program (MEDCAP) on January 2, 2003. The MEDCAP offers medical and eye assistance as well as humanitarian aid that consists of food, clothing, and school supplies.

Figure 2.2 Regional Map of Afghanistan

Source: Wiley C. Thompson

Figure 2.3 Local Map of Afghanistan

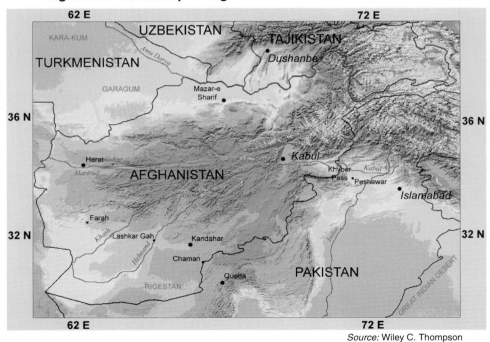

Source: Wiley C. Thompson

Source: Jeremy T. Lock

A couple rides through the streets of Kabul. Bicycles are a popular means of transportation in both urban and rural areas.

A young boy enjoys having his picture taken in the village of Baba Quachar, March 12, 2003. Young children gathered for the grand opening of the new Baba Quachar Elementary School. The school was built by an Afghan contractor and funded by the Coalition Joint Civil Military Operations Task Force (CJCMOTF).

Source: Jeremy Colvin

Figure 3.1 Map of Afghanistan

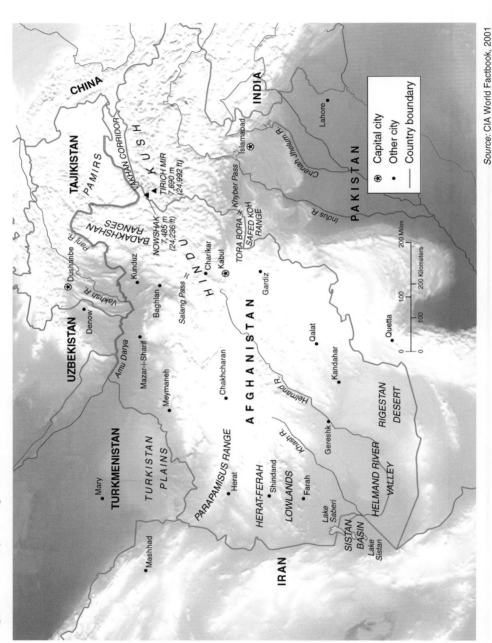

Source: CIA World Factbook, 2001

Figure 12.6   Afghanistan's Summer Temperature Extremes

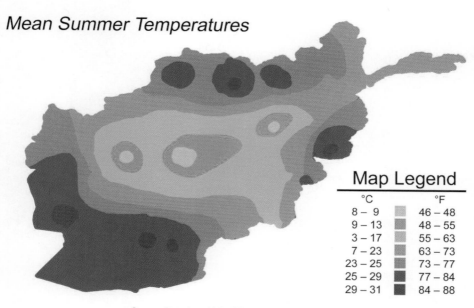

Mean Summer Temperatures

Map Legend

°C		°F
8 – 9		46 – 48
9 – 13		48 – 55
3 – 17		55 – 63
7 – 23		63 – 73
23 – 25		73 – 77
25 – 29		77 – 84
29 – 31		84 – 88

Source: Data from United Nations, Food and Agricultural Organization, 2001

Figure 12.6   Afghanistan's Winter Temperature Extremes

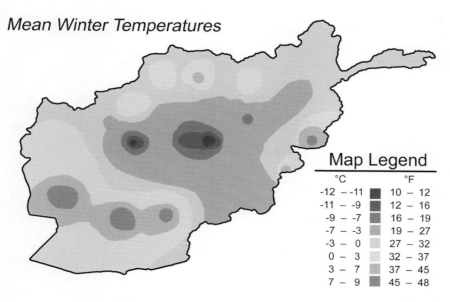

Mean Winter Temperatures

Map Legend

°C		°F
-12 – -11		10 – 12
-11 – -9		12 – 16
-9 – -7		16 – 19
-7 – -3		19 – 27
-3 – 0		27 – 32
0 – 3		32 – 37
3 – 7		37 – 45
7 – 9		45 – 48

Source: Data from United Nations, Food and Agricultural Organization, 2001

Source: John Wiegand

Irrigated agricultural fields of the Panjsher Valley present a stark contrast to the otherwise barren landscape.

Source: Eugene Palka

After serving as the head of the interim government, Hamid Karzai was elected as the country's first president amid hopes of peace and economic development.

In Bamyan, this family prepares to go home after receiving their rations of humanitarian aid from U.S. Special Forces soldiers in April 2002. Each person received one bag of wheat, rice, beans, a sweater (provided by the SF wives club) and a handout that tells of the dangers of land mines.

Source: Ronald A. Mitchell

Source: Andres Rodriguez

For this young Tajik girl, relief from war, improved conditions for women, access to education, and humanitarian assistance provide hope for the future.

Source: Cecilio Ricardo

A rider tries to hold onto the calf carcass while racing toward the winner circle during a heated game of *buzkashi (buz,* "horns" and *kashi,* "to pull") in Mazar-i-Sharif. *Buzkashi* is a popular sport that takes place on a vast field, where men on horseback struggle to pull an 80-pound calf carcass around a flag for 1 point or into a chalk-white circle for 2 points.

Source: Andres Rodriguez

In a picture that reveals hope and guarded optimism, young girls in the village of Charikar attend school for the first time since the Taliban seized power.

A young girl in Mazar-i-Sharif holds up a leaflet warning people not to pick up unexploded ordinance in the area. The U.S. personnel in Mazar-i-Sharif have handed out over 7,000 leaflets such as this one to help protect and warn the people of these dangers.

Source: Cecilio Ricardo

Source: Cherie A. Thurlby

Military personnel bargain with vendors outside Bagram airbase. The vendors sell an assortment of crafts, carpets, clothing, and jewelry on a periodic basis.

Source: Milton H. Robinson

Fresh meat hangs in the doorway of a Charikar establishment, ready for purchase by the townspeople.

Source: John Wiegand

A young boy departs his village and takes a leisurely stroll along the road toward the local market.

A girl is left homeless after flooding by the Panshir and Ghorband Rivers in Parwan Province. U.S. Civil Affairs Battalions provided humanitarian assistance to her and other citizens of the province in April 2003.

Source: Vernell Hall

Source: Eric E. Hughes

Kahrkia Kalay, a small village outside the city of Khowst, is surrounded by the distinctive vegetation of the country's southeast border area.

In the village of Bakhshi Kyhal, a young boy carries an armful of candy that has been passed out by U.S. airmen participating in the Adopt-A-Village Program. Military units help to improve the way of life in a village of their choice by providing such items as clothing, shoes, food, and school supplies.

Source: Milton H. Robinson

Source: Cherie A. Thurlby

The sun rises over the mountains surrounding Bagram airbase.

Source: Andres Rodriguez

Young boys mingle in a school yard while a law-enforcement official patrols the area.

Source: Eugene Palka

A patron departs the one and only commercial store serving the Bagram area. The store and its immediate surroundings bear the visible scars of more than twenty years of warfare in the area.

Source: Eugene Palka

A small irrigated rice paddy field is nestled among the village walls in Charikar.

Source: Jeremy T. Lock

A burial mound rests in the middle of the village of Markhanai (pop. 2,000), which lies in a valley in the Tora Bora region of Afghanistan.

Source: Cherie A. Thurlby

Contract workers compete for a wide range of jobs and services at all military operating bases throughout the country.

2

Location

Wiley C. Thompson

Key Points

- Afghanistan is a landlocked state that shares its borders with six other countries.
- The country occupies a location at the crossroads of Central Asia.
- Slightly smaller than the state of Texas, Afghanistan covers approximately 652,000 sq km (251,672 sq mi).

T he term *location*, which refers to a position or point in space, is fundamental to geographers, who are curious about the spatial distribution of the world's human and physical phenomena. Knowing the location of something enables one to place it within a larger context or framework. To locate is to relate; but perhaps more important, the location of a place serves as the initial point of departure for more detailed geographic inquiry about that place.

ABSOLUTE LOCATION

One of the keys to understanding the importance and complexity of Afghanistan and its role in current events lies in its location. As geographers seek to answer the question of "where?" in their regional analysis, they examine the concept of location in two ways. The first is to examine a region in terms of its *absolute location*. The latter is a fundamental geographic concept that refers to the exact position of a place on the surface of the earth. One often sees absolute location expressed in terms of latitude and longitude or some other coordinate system. The geographic center of Afghanistan lies at approximately 33°N latitude and 65°E longitude (Figure 2.1).

The north-south extents of Afghanistan's borders run from 38°26′N in the northeastern corner to 29°23′N along its southern border. A proruption into the Hindu Kush range makes Afghanistan appear to be deceptively wide as its west-east extents run from 60°34′E to 74°53′E, a distance of 1,300 km, equiv-

Figure 2.1 Location of Afghanistan, Global Scale (33°N, 65°E)

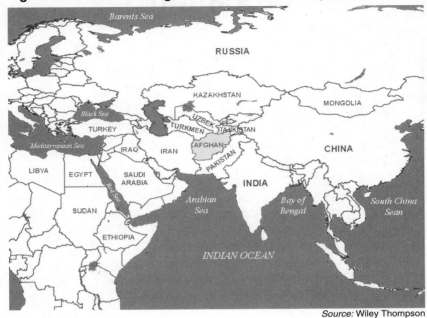

Source: Wiley Thompson

alent to the distance between New York City and Fort Lauderdale, Florida. The entire country is located within the + 4:30 time zone.

What does Afghanistan's absolute location signify for an American traveler? It means that to reach the capital, Kabul, a person leaving from New York City must cover 10,862 km (6,745 mi). A nonstop flight would take approximately 13 hours. From the West Coast, a nonstop flight from Los Angeles to Kabul would cover 12,362 km (7,677 mi) in 12 1/2 hours.

RELATIVE LOCATION

Another method by which geographers answer the question of "where?" is through the concept of *relative location*. This describes the location of a place relative to the position of other places or phenomena. The relative locational perspective is affected by distance and accessibility to other resources and influences within the larger region or realm. The relative location of a region is a key element in the geographer's analysis of the historical, cultural, political, and economic geography of that region. A cursory introduction to Afghanistan's relative location is given here; a critical analysis follows in later chapters.

Afghanistan is a Central Asian, landlocked country that is bordered by Turkmenistan, Uzbekistan, and Tajikistan to the north, China to the east, Pakistan to the east and south, and Iran to the west (Figure 2.2 [C-2]). The country

Source: Wilson Guthrie

Mountain ranges encircle Bagram Airbase in Parvan Province.

is only slightly smaller in size than the state of Texas, but finds itself at the crossroads of western Asia. It has played the role of a buffer state between the former Soviet and British Empires, and although Afghanistan has been an independent country since 1919, its proximity to the former empires has left a lasting imprint on its landscape and people.

With its high, rugged mountains in the east and its arid plains in the north and southwest, the country has played the role of a religious buffer among a number of religious traditions: the practices of Shia and Sunni Islam, Sikhism, Hinduism, and Buddhism converge in this region (Figure 2.3 [C-2]). While the Taliban Party adhered to a fundamentalist form of Islam, the proximity of other religious groups or variations of Islam undoubtedly proved to be as much of a centrifugal or dividing force within the country, as it was centripetal or unifying phenomenon. Likewise, Afghanistan's relative location to other distinct and diverse cultures in the region is apparent in the many ethnicities and languages found within its political borders.

SUMMARY

The concepts of absolute and relative location are key to the geographer's analysis and understanding of a region. These concepts are the starting points for discovery and can be applied to various themes or geographic subfields as the authors focus on how the people have interacted with their neighbors and the surrounding environment. It is through this framework that the geographer can better explain the physical and human geography of Afghanistan.

3

Geomorphology

Matthew R. Sampson

Key Points

- Afghanistan's physical relief is dominated by the rugged Hindu Kush Mountains.

- The rugged terrain restricts overland travel and canalizes movement through the mountain passes.

- Earthquakes are common in the northeast provinces and recent large-magnitude events have caused thousands of deaths.

Geomorphology is the study of landforms and the processes that shape them. It entails understanding the terrain, or the lay of the land, and how it got that way. The geomorphic processes that shape the earth's surface include weathering and erosion. The former breaks down surface materials either by physical or chemical means. The latter refers to the movement of weathered surface material by wave action, running water, blowing wind, or glacial ice. The result is a dynamic landscape that is continually being reshaped by the forces of nature. Depending on the dominant geomorphic forces at work, distinct physical landscapes will result. If tectonic forces are dominant, for example, they would cause a mountainous landscape. Geographers often categorize these distinct landscapes into geographic regions.

GEOGRAPHIC REGIONS

Afghanistan is dominated by rugged, mountainous terrain. The massive Hindu Kush Mountains form a barrier between the northern provinces and the rest of the country. This mountain range divides Afghanistan into three distinct geographic regions: the central highlands, the northern plains, and the southwestern plateau [C-4]("Land and Resources," *Afghanistan Online*, 2001).

Geomorphology

Source: Andres Rodriguez

The rugged Hindu Kush Mountains dominate the physical landscape.

The Central Highlands

The central highlands comprise about 70 percent of Afghanistan. This region consists primarily of the Hindu Kush, which occupy the center of the country. It is a rugged, snowbound highland that is one of the most impenetrable regions in the world (English, 1984). The Hindu Kush range extends for about 1,000 km (625 mi) in a southwesterly direction from the Vakhan Corridor in the northeast almost to the border with Iran in the west. From the Hindu Kush, other lower ranges radiate in all directions.

The Hindu Kush form the western extremity of the Himalaya and consists primarily of granites and schists that were probably uplifted during the Tertiary period (66–2 million years ago). Within the system there are also areas marked by the overthrust of Cretaceous limestones on Cenezoic shales and clays ("Hindu Kush," *Microsoft® Encarta® Online Encyclopedia*, 2001). The average elevation of this mountainous region is 2,700 m (8,775 ft.) and the highest peak, Nowshak, in the northeast, reaches 7,485 m (24,326 ft.). Small glaciers and year-round snowfields are common (*Lonely Planet*, 2001). The Hindu Kush extends southward from the Pamirs as a wide plateau, dotted with small glacial lakes. The mountains increase in elevation to the southwest, boasting peaks such as the Tirich Mir, 7,690 m (24,992 ft.) in height, in Pakistan. Many other peaks rise to more than 6,100 m (19,825 ft.) in elevation ("Hindu Kush," *Encarta®*, 2001).

Other major mountain ranges within Afghanistan include the Pamirs in the upper northeast of the Vakhan Corridor, the Badakhshan Ranges in the northeast, the Parapamisus Range in the north, and the Safed Koh Range, which forms part of the frontier between Afghanistan and Pakistan.

The Northern Plains

The most fertile part of Afghanistan, this region occupies about 15 percent of the country. It consists of foothills and plains through which the Amu Darya flows. The average elevation is about 600 m (1,950 ft.) above sea level ("Land and Resources," *Afghanistan Online*, 2001).

The Southwestern Plateau

This region of Afghanistan is made up of high plateaus and sandy deserts. It is essentially the lowland area of Afghanistan and includes the Turkistan Plains, the Herat-Ferah Lowlands of the extreme northwest, the Sistan Basin and Helmand River valley of the southwest, and the Rigestan Desert of the south ("Afghanistan," *Encarta®*, 2001). The soil here is very infertile, except along the rivers in the southwest. The average altitude of this area is about 900 m (2,925 ft.) above sea level. Sandstorms are common in the deserts and arid plains of this region ("Land and Resources," *Encarta®*, 2001).

KEY TERRAIN FEATURES

Given the rugged nature of its terrain, the harsh climate, and the challenges to overland movement, several types of terrain features warrant further discussion. Mountain passes, caves, an earthquake-generating plate boundary, and several exotic rivers serve as distinguishing features of the physical environment, and each has an impact on human settlement patterns. Passes facilitate movement through mountainous terrain, and are as important today as they were 3,000 years ago. Caves are an integral part of the country's subsurface terrain and continue to be a part of the current habitat. The transform plate boundary that bisects the eastern portion of the country makes this region one of the most earthquake-prone areas of the world. Rivers provide the means for life in arid parts of the country.

Passes

Because Afghanistan has many high mountains, the passes through them have been of profound importance in the history of the country. In the 320s B.C.E., Alexander the Great invaded the country through the Kushan Pass, about 4,370 m (14,200 ft.), in the west and exited it to the east through the low Khyber Pass, 1,072 m (3,517 ft.), to invade India. The famous Salang Pass 3,880 m (3,517 ft.) and its Soviet-built tunnel in the central Hindu Kush was one of the main routes the Soviets used to invade Afghanistan in 1979 ("Afghanistan," *Encarta®*, 2001).

Geomorphology

Source: Albert Eaddy

Caves and tunnels are abundant within Afghanistan.

The Khyber Pass is the most important pass connecting Afghanistan and Pakistan. It is controlled by Pakistan and winds northwest through the Safed Koh near Peshawar, Pakistan, for about 48 km (30 mi) to Kabul, Afghanistan, varying in width from 5 to 137 m (16 to 445 ft.). The mountains on either side can be climbed only in a few places because of steep cliffs that surround the pass ("Khyber Pass," *Encarta®*, 2001).

Caves

Underground formations that could hide and shelter terrorists became a topic of major interest during the recent conflict in Afghanistan. Usually caves form in limestone, which is susceptible to dissolution by flowing groundwater. But limestone is limited to only two small areas of Afghanistan: the high tablelands north and west of Kabul, and a lower, desert massif northeast of Kandahar (Marvel, 2001). Man-made caves and tunnels are abundant within Afghanistan. Some are centuries old and were dug for various uses, such as habitation, religious shrines, mineral extraction, and as irrigation tunnels known as *karez* (Schindler, 2002). Common in the Middle East, karez are underground tunnels that carry water downhill from wells, which are often located on alluvial fans. Sometimes karez are also used as hideouts or to store contraband.

The Tora Bora area near the border with Pakistan has become well known for its network of caves and tunnels. These are unique because the dominant lithology is metamorphic gneiss and schist. Originally excavated by the Afghan mujahedin during the Russian invasion, the Tora Bora tunnels were

Figure 3.2 Major Earthquakes Since February 1998

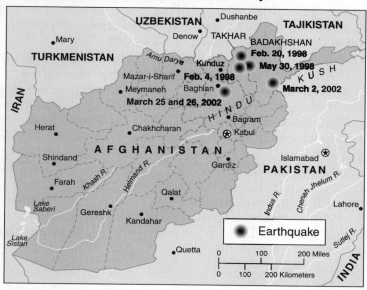

Source: "Impact of the Earthquake," 1998

apparently expanded by al-Qaeda using hard-rock mining techniques. Reinforced with steel and concrete, the crude caves and tunnels were converted into sophisticated bunkers with multiple entries, exits, and air holes (Shroder, 2001).

Plate Boundaries and Earthquakes

The northeast portion of Afghanistan is one of the most seismically active areas in the world. In terms of plate tectonics, this is an area where a transform boundary exists between the Eurasian plate and the Indo-Australian plate. For the past 65 million years, the Indo-Australian plate has been moving in a northeasterly direction converging with the Eurasian plate, resulting in the formation of the Himalayas. These two plates slide past one another in the area of Afghanistan and Pakistan, which produces frequent earthquakes (Hudson & Espenshade, 2000). Recent earthquakes include one in Takhar province on 4 February 1998, about 50 km from the border with Tajikistan. The earthquake (magnitude 6.1 on the Richter scale) affected 28 villages, killed about 4,000 people, and left another 20,000 without shelter ("Thousands Reported Killed in Afghanistan Quake," 1998). On 20 February 1998 another strong earthquake occurred in northeastern Afghanistan with a magnitude of 6.4 on the Richter scale. No injuries or deaths were reported (OCHA, 1998). Just three months later, on 30 May, a magnitude 6.9 earthquake occurred along the boundary of Badakshan and Takhar provinces, again in northeastern Afghanistan (Figure 3.2). Approximately 4,000 people were killed, and as many as 70,000 people in 70 villages were affected ("Impact of the Earthquake," 1998).

In the mountainous areas of Afghanistan, earthquakes can trigger devastating landslides. Landslides of mud, made worse by heavy rain in the days preceding the 30 May earthquake, caused immense destruction, making road access impossible in many areas. Simple rural dwellings perched on the mountainsides were simply swept away. The landslides also destroyed water sources with the collapse of wells and springs ("Impact of the Earthquake," 1998).

During the spring of 2002, three major earthquakes occurred in the area north of Kabul. The first occurred on 3 March 2002, with the epicenter located less than 150 mi from Bagram, the major base of operations for Coalition forces. More than 100 people were killed. Two additional quakes occurred on 25 and 26 March within only 50 mi of Bagram. One measured only 5.9 on the Richter scale, but the shallow nature of the quake produced extensive damage and an estimated 1,800 dead.

Rivers

Many of Afghanistan's major rivers are fed by mountain streams. Consequently, the greatest stream flow occurs in the late spring and early summer with the arrival of snowmelt from the mountains, bringing with it the danger of flooding. The Amu Darya is the largest river of Central Asia and measures 2,540 km (1,588 mi) in length. The Amu Darya's main tributaries are the Panj and Vakhsh Rivers, both of which rise in the Pamirs. The Panj forms part of the boundary between Tajikistan and Afghanistan, and the Vakhsh flows through southwest Tajikistan to join the Amu Darya at the Afghan border. The Amu Darya follows a northwest course between Tajikistan and Afghanistan, continues northwest between Turkmenistan and Uzbekistan, and then flows north through Uzbekistan into the Aral Sea. Over the centuries the river has shifted its course several times. In the third and fourth millennia B.C.E., the Amu Darya flowed westward from the Khorezm Oasis into Lake Sarykamysh and from there to the Caspian Sea. From the seventeenth century until the 1980s the Amu Darya emptied exclusively into the Aral Sea, except during periods of intense flooding, when overflows went into Lake Sarykamysh ("Amu Darya," *Encarta®*, 2001).

During the 1980s several years passed in which little or no water reached the Aral Sea from the Amu Darya. The largest single cause of the decline in the river's water level is the Garagum Canal, the longest canal in the former Soviet Union and one of the longest in the world. Near the town of Oba, Turkmenistan, the canal diverts water from the river at the rate of about 12 cu km per year—about one-ninth of all the water diverted in the Aral Sea basin. Reduced water flow has restricted water transportation on the Amu Darya, which was once navigable for light draft vessels over nearly half its length ("Amu Darya," *Encarta®*,

Source: Eugene Palka

A village of the Panjsher Valley is dwarfed by mountains in the background.

2001). Still, it is the only navigable river in Afghanistan, although ferryboats can cross the deeper areas of other rivers.

Both the Harirud and Helmand Rivers rise in central Afghanistan. The Harirud flows west and northwest forming part of the border with Iran. The Helmand also flows toward Iran but in a southwest direction. Its importance for irrigation and agriculture has declined, as it has become increasingly salty over the years. Interestingly, most of Afghanistan's rivers never reach the sea. Instead they drain into inland seas or salt flats. The one exception is the Kabul River, which eventually reaches the Indian Ocean via the Indus River ("Afghanistan," *Encarta®*, 2001).

SUMMARY

Afghanistan is best described as a rugged landscape with significant vertical relief covering an extensive portion of the country. Underlying processes continue to shape the earth's surface in the region and produce unpredictable, if not catastrophic, changes to the physical landscape. These internal and external processes continue to contribute to regional isolation among the country's inhabitants and pose a formidable barrier to anyone who attempts long-distance travel within the country.

4

Climatology

Richard P. Pannell

Key Points

- Afghanistan's climate is characterized by extremes.
- To understand the country's climate, one must appreciate the impacts of key climate controls such as latitude, continentality, and topography.
- Significant climate hazards include drought, high winds, dust storms, and heavy snowfall.

Climatology is the subfield of geography that examines the long-term conditions of the atmosphere and the interactions between the atmosphere and the earth's surface. Climate profoundly influences a host of environmental processes such as vegetative growth, soil formation, watershed hydrology, and geomorphic denudation. Climate also influences human activities in a number of ways ranging from the development of agricultural practices to the conduct of military operations. The term *climate* is often referred to as the "average weather" of a location, but this is overly simplistic and misleading. Climatology can be used to examine the spatial distribution of atmospheric extremes, atmospheric teleconnections, and atmospheric variability in addition to organizing the patterns of average conditions. Depending on the spatial scale considered or the timeframe used, a variety of different patterns emerge in the analysis of any region. As the spatial scope narrows and as the timeframe shortens, a host of factors internal to the climate system begin to play an increasingly important role in developing a clear analysis of "average" atmospheric conditions. Factors such as topography, vegetation, regional pressure variability and even *El Niño* phases may play an important part in determining the climate for a particular season or particular year. This is especially important in our analysis of Afghanistan.

While the average conditions of Afghanistan at first glance appear to be associated with an arid desert or semiarid steppe climate, the reality is much different due to macro scale pressure oscillations associated with the Asian monsoon, the massive Hindu Kush Mountain range that bisects the country,

Figure 4.1 Regional Köppen Climate Classification Applied to Afghanistan

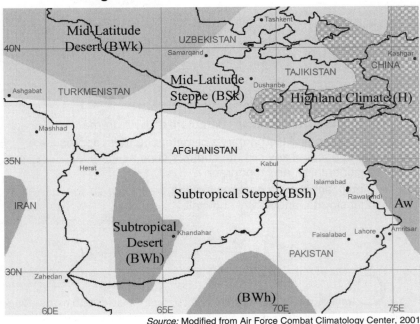

Source: Modified from Air Force Combat Climatology Center, 2001

and a drought since 1999 that lasted for almost three years. In order to develop a working climatology of Afghanistan, it is useful to first examine the mean climate conditions based on the data available. Following the regional climatology, climate controls are assessed, and the influence of other factors, such as local topography, altitude, and wind patterns are addressed to develop a clearer picture of the local climate in several areas of the country. Finally, climate variability is examined, particularly in light of the factors influencing the potential for continued drought.

OVERVIEW

Afghanistan sits astride the 35th parallel, making it similar in latitude to the states of New Mexico and Arizona. In some respects the climate of Afghanistan is similar as well, exhibiting subtropical to midlatitude steppe in most areas, while others are drier subtropical deserts. For the purposes of our climate analysis, Köppen's climate classification system as modified by Trewartha is used to describe climate types (GWA, 2000). Figure 4.1 shows the regional Köppen climate classification. Like the desert Southwest United States, Afghanistan has hot summers and cool winters, although winters in the Afghani highlands are more extreme. Summertime high temperatures frequently exceed 38°C (100°F) particularly in the southwest regions, while wintertime lows can reach below –25°C (–12°F) (AFCCC, 1995). The presence of high mountains influences local climates by decreasing temperatures at high altitudes and affecting the precipitation regimes. The climates of Afghanistan and the Southwest United States are very different, however. Unlike the Southwest United States, the majority of precipitation occurs in the winter and early spring in Afghanistan, where conditions in many areas such as Kabul are cloudy, foggy, and snowy. Periods of increased precipitation often result in flash flooding as ephemeral stream channels are quickly filled by thunderstorm activity or monsoon-like rains (AFCCC, 2001). These differences in climate can be attributed to a host of climate controls that operate differently in Afghanistan than in the Southwest United States.

CLIMATE CONTROLS

Climate controls influence a variety of climatic variables such as temperature, temperature range, precipitation, and wind. These controls provide insight into the critical factors that determine the climate and weather regimes of Afghanistan. Some climatic factors such as ocean currents play very little role in determining Afghanistan's climate due to its location in the interior of the Asian continent. Other controls such as latitude, *continentality*, and topography are central to explaining the climate of the country. In examining these climate controls, climographs such as those of Kandahar and Faizabad are used (Figure 4.2).

The amount of incoming solar radiation, or *insolation*, that Afghanistan receives is primarily a function of its latitude. Between 29° and 38°N latitude, Afghanistan lies just north of the subtropics. As a result, the country receives relatively high amounts of insolation, particularly in summer when the northern hemisphere is tilted toward the sun, and periods of daylight are longer. This brings warm temperatures with daily maximum temperatures across the country often exceeding 38°C (100°F). Additionally, this high level of insolation creates high potential evapotranspiration rates, affecting the availability of soil moisture. Winter insolation is considerably less as the sun is lower on the horizon and day lengths are shorter, and therefore temperatures are cooler.

In addition to the variability of insolation in Afghanistan, the location of the country within the Asian landmass results in broad annual and daily temperature ranges. The rapid heating and cooling of the land surface due to its relatively low specific heat produces a controlling factor known as *continentality* (McKnight 2002). As such, periods of heating, such as summer or the middle of the day, can witness rapid temperature increases, while periods of reduced insolation, such as winter or nighttime, experience rapid cooling. This situation creates climates of extreme temperatures from summer to winter and even from

Climatology

Figure 4.2 Köppen Climographs for Kandahar and Faizabad

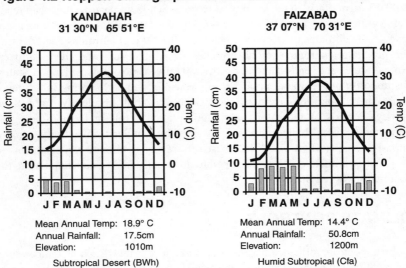

KANDAHAR
31 30°N 65 51°E

FAIZABAD
37 07°N 70 31°E

Mean Annual Temp: 18.9° C
Annual Rainfall: 17.5cm
Elevation: 1010m

Subtropical Desert (BWh)

Mean Annual Temp: 14.4° C
Annual Rainfall: 50.8cm
Elevation: 1200m

Humid Subtropical (Cfa)

Source: Data from Air Force Combat Climatology Center, 1995

day to night. Figure 4.2 shows broad annual temperature ranges of over 20°C (36°F) for both stations.

The temperatures in Afghanistan vary widely due to the variability in topography. Altitude has a dramatic effect on temperatures. Air temperatures decrease as elevation increases as a function of the environmental lapse rate, which averages about 6.5°C per 1,000 m (11°F per 3,250 ft.) of elevation. With much of Afghanistan dominated by the Hindu Kush Mountains, altitude must be considered carefully. Peaks exceed 5,000 m (16, 250 ft.), and consequently these mountains have temperatures that may vary by 10 to 20°C (18 to 36°F) over relatively short horizontal distances. Such variability in temperatures makes it very difficult to characterize broad regions as having uniform climate types, and therefore, within the Köppen climate classification, the use of undifferentiated highlands climate (type H) is used to designate these areas.

Precipitation regimes in Afghanistan are largely controlled by surface pressure changes and the effect of *orographic* precipitation, that is, precipitation that results when air ascending over a topographical barrier cools to its dew point). Winters are influenced by the Siberian high pressure system, which spreads cold, dry continental air outward in all directions and pushes the subtropical jet stream south of the Himalayas. During this period prevailing winds are typically from the northwest or north, resulting in the possibility of storms from the western Mediterranean tracking across Afghanistan every few days. As a result, the potential for precipitation is greatest in the winter and early spring and is often in the form of snow. As the Asian landmass heats up in

Climatology

Figure 4.3 Köppen Climographs for Salang Tunnel and Chakhcharan

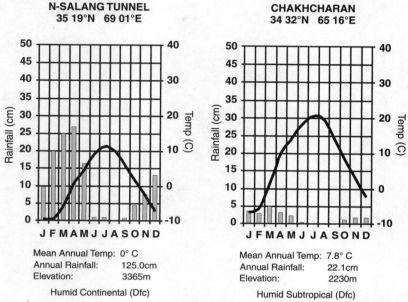

N-SALANG TUNNEL
35 19°N 69 01°E

CHAKHCHARAN
34 32°N 65 16°E

Mean Annual Temp: 0° C
Annual Rainfall: 125.0cm
Elevation: 3365m

Humid Continental (Dfc)

Mean Annual Temp: 7.8° C
Annual Rainfall: 22.1cm
Elevation: 2230m

Humid Subtropical (Dfc)

Source: Data from Air Force Combat Climatology Center, 1995

summer, the subtropical jet shifts north of the Himalayas, bringing the area of Afghanistan warm dry air from the north and northeast, with little or no precipitation. Typically blocked by mountains, the southwest monsoon that affects India and Pakistan can on occasion cause thunderstorms in eastern Afghanistan for three or four days at time (AFCCC, 1995).

The influence of topographic barriers has a dramatic effect on precipitation regimes throughout the country. Orographic precipitation on the windward side of mountain ranges is common as moist air is forced upward and cooled adiabatically. In the winter this is common, particularly as low-pressure systems track from the northwest over the Hindu Kush. For example, the Salang Tunnel region amid the highest peaks of Afghanistan averages almost 130 cm (51 in) of precipitation a year (Figure 4.3). By contrast, on the leeward side of mountain ranges air descends and warms adiabatically resulting in a "rainshadow" effect where the air is warmer and much drier. The city of Chakhcharan in central Afghanistan is on the leeward side of the Paropamisus Range and receives less than 25 cm (10 in) of precipitation (AFCCC, 1995).

By examining the climate in Afghanistan in terms of controlling factors, it becomes apparent that the generalizations of climate types in Figure 4.1 are inadequate. These climate controls and the climate data from existing stations in Afghanistan provide a better understanding of climatic regions.

Figure 4.4 Regional Climate Patterns Based on Station Data

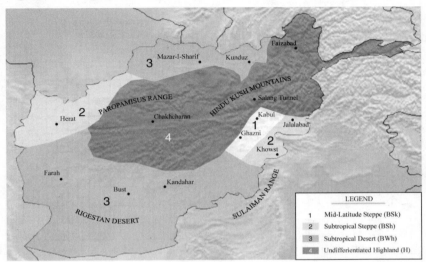

Source: Data from Air Force Combat Climatology Center, 1995

CLIMATE REGIONS

Analysis of climate data from multiple stations creates a much different picture of the regional climate. By examining the local climate classifications of the 13 climate stations in Figure 4.4 and considering the climate controls that influence the country, a more appropriate climate picture emerges.

Clearly the Hindu Kush Mountain range plays a dominant role in determining local climate. The analysis of climate stations at Chakhcharan, Salang Tunnel, and Faizabad demonstrates the variety of climates that can be found within these highland areas. Both Chakhcharan and the Salang Tunnel are humid continental (Type D) climates, while Faizabad is actually a humid subtropical (Type C) climate. The increased precipitation due to orographic lifting and the cooler temperatures create very unexpected climates for the region. Also, variability in altitude, slope aspect, and prevailing winds make the central highlands of Afghanistan difficult to characterize. As such, the latter are labeled as Undifferentiated Highlands (H). Extreme conditions are the norm here and severe events such as heavy snowfall ensure that the roads and trails are impassable for periods of weeks or months. It is common for the high passes to be closed from November through late March and snowpack is permanent for elevations above 3,655 m (11,900 ft.) (AFCCC, 1995).

The northern region around Mazar-i-Sharif is classified as a subtropical desert (BWh) climate based on average temperatures above freezing and little precipitation. The precipitation that does occur peaks in the early spring and is highly variable. Again, this is dependent on the strength of the Siberian high. Surrounding the region are areas of variable rainfall that may be classified as subtropical steppes (BSh) in some cases. Kunduz to the east is one of these areas that receives more rainfall. With the drought of 1999–2001, however, the current climate is likely to be more desert-like. To the southwest conditions are wetter due to the increased frequency of low-pressure disturbances. The area around Herat is a subtropical steppe (BSh) climate, but here again, precipitation may be highly variable.

The southern region of the Rigestan Desert is the largest expanse of desert in the country. The area around Kandahar, Bust, and Farah is dominated by the subtropical high throughout the year and is extremely dry. The mountain ranges north of this region protect it in the winter from the effects of the northeast monsoon resulting in dry winters, although temperatures may fall below freezing. Summers are extremely hot with temperatures exceeding 45°C (112°F) in July (AFCCC 1995).

Finally, in the eastern steppe areas around Kabul and Jalalabad, conditions are cooler and moister, particularly in winter. Near the Pakistani border data indicate that both Jalalabad and Khowst are subtropical steppe (BSh) climates, but there is a climate gradient indicated westward as elevations increase. Both Kabul and Ghazni are at considerably higher elevations and average monthly temperatures in winter are below freezing resulting in the classification of midlatitude steppes (BSk) in these areas. All of these areas are subject to migratory low-pressure systems in winter and are occasionally cloudy and rainy during these periods.

CLIMATIC ANOMALIES

Up to this point, the spatial patterns of climate types in Afghanistan have been examined using climate data, climate controls, and narrative climatologies from the Air Force Combat Control Center. None of these resources, however, can completely account for the presence of the long-term drought that has gripped the country since January 1999. Such a severe drought represents a significant anomalous pattern from the expected climate norm. The cause of these droughts is closely related to the lack of precipitation in the winter and spring, which in turn is linked to the strength of the Siberian high pressure. The stronger the Siberian high, the more likely westerly low-pressure systems will be blocked from reaching Afghanistan. In addition to the blocking of low-pressure systems, surface temperatures have been anomalously high in Afghanistan over the last year. This leads to increased potential evapotranspiration, thereby exacerbating drought conditions. Typical evapotranspiration rates exceed available moisture by an order of magnitude. The ultimate cause of these drought conditions remains indeterminate. Linkages to known teleconnections such as *El Niño* cannot be statistically correlated. To date the effects of these droughts have been severe. The FAO has issued multiple

Climatology

Source: Andres Rodriguez

The early morning sunlight over Bagram suggests another hot and dusty day ahead.

famine alerts over the last two years on Afghanistan and estimates that about 6 million people are vulnerable to the effects of crop failure and famine as a result of the prolonged drought (FAO 2001).

SUMMARY

The climate of Afghanistan is complex and is not fully understood. Attempting to characterize it regionally is difficult due to the absence of data and the high-altitude mountains that cover much of the country. Climate controls such as altitude and continentality significantly affect the temperature regimes, while pressure systems and topographic barriers influence precipitation patterns. The overall result is a climate of extremes: hot summers and bitter winters, arid deserts and snowpacked highlands. Additionally, the potential for severe anomalous conditions such as drought are being felt today.

5

Biogeography

Peter G. Anderson

Key Points

- Limited forest development occurs in the eastern part of the central highlands.
- Western Afghanistan is subdivided into the southern deserts and northern grasslands.
- Natural vegetation is minimal due to aridity and human activity, and environmental degradation is extensive.

B iogeography is the study of the distribution of plants and animals— where these biotic entities occur, and why they occur at these locations. This field of study utilizes information derived from many other disciplines and subfields, such as meteorology, climatology, geomorphology, physiography, botany, zoology, ecology, and land use and resource management. The biogeographic scale of study normally is regional to continental and global. The biogeographic study of Afghanistan is a regional scale study, consisting of southern deserts, northern grasslands, and a central highlands area.

Studies related to Afghanistan's plant and animal mosaic are few. Saba (2001) states that there is a lack of vegetation and other environmental information pertaining to Afghanistan. Without a database from which to draw conclusions, only general inferences may be made about the country's biogeography. As such, this chapter will characterize the biogeography of Afghanistan based on general vegetation characteristics. This description refers to a natural vegetation cover, although the natural vegetation for much of the world has been altered during the past thousand years due to natural processes and human activity. Afghanistan is no exception.

TYPES OF VEGETATION

Afghanistan is a country of three primary climatic, physiographic, and vegetation regions: the central highlands, the southern plateau and desert, and the northern plains and grasslands. It is a country of approximately 652,500 sq km

Biogeography

Figure 5.1 Vegetation Patterns in Afghanistan

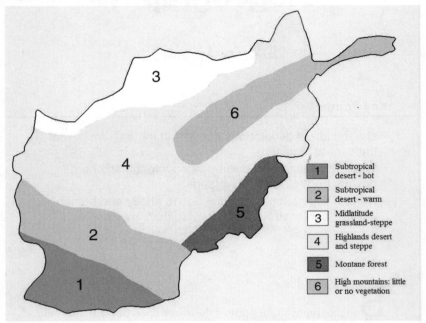

Legend:
1. Subtropical desert - hot
2. Subtropical desert - warm
3. Midlatitude grassland-steppe
4. Highlands desert and steppe
5. Montane forest
6. High mountains: little or no vegetation

Source: Peter Anderson, data based on Saba (2001)

(251,825 sq mi) (*Encarta®*, 2001), of which 12 percent is arable land, 3 percent is forestland, 46 percent is pastureland, and the remaining 39 percent is termed "other" (CIA 2001). Saba (2001) states that two-thirds of Afghanistan is mountainous and supports little or no vegetation; one-sixth of the country is desert land; and one-sixth of the land is pasture and farmland. Essentially, these land characterizations relate to the central highlands, the desert south, and the northern grasslands, the arable lands. Figure 5.1 portrays the vegetation patterns of Afghanistan.

Central Highlands

The central highlands region of approximately 414,000 sq km (160,000 sq mi) (Saba 2001) consists of high mountains and deep valleys. The elevation of about one-half of Afghanistan is greater than 2,000 m (6,600 ft.) (*Encarta®*, 2001). The climate of this mountainous region may be classified as highlands, since the weather and climate change with increasing elevation. The normal characteristics are warm, dry summers and cold winters.

These conditions, characterized by high elevation that is warm and dry with cold winters, are not conducive to plant growth. Most of the highland

Table 5.1 Major Forest Types, Dominant Species, and Their Elevational Distribution in Afghanistan

Forest type	Dominant species	Occurrence
Temperate Forest Group		
1) Montane Moist Forest	*Pinus, Quercus, Acer, Betula, Alnus, Abies, Picea, Tsuga, Taxus, Machilus, Rhododendron*	1800–3600 m
2) Montane Dry Forest	*Pinus, Quercus, Betula, Picea, Tsuga, Cedrus, Rhododendron, Larix, Juniperus, Acacia, Zizyphus, Caragana, Populus, Salix, Fraxinus*	2100–3600 m
Alpine Forest Group		
1) Sub-Alpine Forest	*Picea, Tsuga, Betula, Juniperus, Rhododendron*	2200–3600 m
2) Alpine Forest	*Tsuga, Betula, Juniperus, Berberis, Rhododendron Caragana, Hippophae, Salix*	2400–3600 m

Source: Bandyopadhyay (1992)

region is classified as subtropical desert and midlatitude grassland-steppe vegetation. Grasses and small, herbaceous forbs that are present tend to be widely spaced. In sheltered sites, along waterways and other moist locations, the vegetation may be more dense and continuous. The primary condition of the highland desert and steppe region, however, is sparse vegetation.

Forests characterize 3 percent of Afghanistan's land and are concentrated in the eastern portion of the central highlands. The location of Afghanistan's forests coincides with the western edge of summer monsoon rain areas, but the amount of rain that falls is minimal compared with the eastern Himalaya. Trees that may be found in Afghanistan's forests (Table 5.1) include: pine, spruce, fir, hemlock, larch, juniper, alder, birch, willow, oak, poplar, ash, rhododendron, wild hazelnut, almond, and pistachio (Bandyopadhyay, 1992; *Encarta®*, 2001). Common shrubs are rose, honeysuckle, hawthorn, current, and gooseberry (Arianae, 2001).

Lower treeline or forest development in Afghanistan occurs at an altitude between 1,678 and 2,196 m (5,500 and 7,200 ft.), whereas the upper treeline occurs higher than 3,050 m (10,000 ft.) (Arianae, 2001). Cedar, juniper, and oak are common tree species of the lower treeline and woodlands (open forest of arid environments) and fir is a common species of the upper treeline.

Southern Deserts

Within the southern desert region, the average elevation of the southern plateau area is about 915 m (3,000 ft.) (Saba, 2001). This 129,500 sq km (50,000 sq mi) region is an area of poor soils, with *Aridisols* being most common. The latter is a soil order typically found in a dry environment where because of inadequate soil moisture the soluble minerals are not removed from the soil. The soil is generally sandy and lacking in organic matter. It is classified as a subtropical desert, and it is hot during summer and mild during winter (*Encarta®*, 2001). Desert conditions are characterized by minimal precipitation and soil moisture deficits.

Afghanistan's southern desert is an area of sparse vegetation and human settlement. Plants common to this area are: camel thorn, locoweed, spiny rest-harrow, mimosa, and wormwood, a variety of sage (*Encarta®*, 2001). Many of the plants found in desert environments are xerophytes or ephemerals. Xero-phytic plants have adaptations that allow the plant to survive during times of little or no soil moisture. These adaptations include the loss of leaves, leathery leaves, stems that store moisture, small size, and deep root systems. Ephem-erals survive from year to year as dormant root mass or as a seed. When suffi-cient soil moisture exists, rapid plant growth and reproduction occurs. These events occur in the southern deserts of Afghanistan following the rainy season of early spring (Arianae, 2001). During this time period, the desert is alive with grasses and small herbaceous forbs in bloom.

In the hottest environments of the southern desert, the landscape is barren and few plants are present. Ephemeral plants are common in this landscape. In the cooler, highland desert landscape, xerophytes are present, as are ephem-erals. In both environments, plants tend to be widely spaced, in relation to plant diameter, with bare ground between individuals.

The Northern Grasslands

The vegetation of the northern plains is classified as midlatitude grassland-steppe. This 40,000 sq mi region includes Afghanistan's most fertile soils and favorable climate (Saba, 2001). Normally, temperature and precipitation re-gimes of this climate favor a positive soil moisture balance. Where soil mois-ture and plant growth are favorable, *Mollisols* (soils characteristic of prairies and grasslands) are found. Where these conditions are not met, Aridisols are present. Plants that are common in a midlatitude steppe are short grasses and herbaceous forbs on moist sites (Mollisols) and bunch grasses and dry shrubs on drier sites (Aridisols).

It is in this region that much of Afghanistan's agriculture production takes place. However, only about one-half of Afghanistan's agricultural lands are currently used as farmland (Saba, 2001). Environmental problems such as desertification, salinization, and chemical contamination have degraded some of the arable lands. Recent warfare has also damaged prime agricultural lands (Saba, 2001, *Encarta®*, 2001), and Afghanis have abandoned some farmlands due to war. Irrigation is needed to grow crops on the drier sites, but most Af-ghanis are too poor to afford modern irrigation technology.

Against a mountain backdrop, time stands still in a village and open fields.

AFGHANISTAN'S WILDLIFE

Beniston (2000) indicates that 123 animal and 460 bird species are known to exist in Afghanistan. Some of the animal species are bears, wolves, foxes, hyenas, jackals, mongoose, wild boar, Marco Polo sheep, urials, ibex, hedgehogs, hares, shrews, bats, and numerous rodents (*Encarta®*, 2001). Among the 123 different animal species are some that are nearing extinction, such as the leopard, snow leopard, goitered gazelle, markor goat, and Bactrian deer (Microsoft 2001). Bird species include partridge, pheasant, quail, vultures, ducks, pelicans, cranes, flamingos, and eagles (Arianae, 2001). Birds that are widely hunted are diminishing in numbers (*Encarta®*, 2001).

SUMMARY

Afghanistan is characterized by a harsh, inhospitable landscape. Very little about this country suggests abundance, and during recent decades, warfare has degraded the quality of life in Afghanistan. Forests have been cut and burned. Agricultural lands are being degraded. Water sources have become polluted. The presence of more than 10 million land mines, however, is considered to be the worst environmental hazard (Saba, 2001). The number of deaths caused by land mines is estimated at between 20 and 30 noncombatants per day (Saba, 2001).

Arianae (2001) suggests that we remember that no serious ecological studies have been conducted in Afghanistan during the past 20 years and we

Biogeography

On April 16, 2003, young Afghan children walk their sheep home in Khar Bolaq, Afghanistan.

don't know the impact that recent warfare has had on the fauna and flora. We do know, however, that the forests of Afghanistan are being used for fuel and building material. They have been negatively influenced during recent warfare. Where the forests have been degraded, erosion has increased. These trends may be expected to continue.

Pasture and agricultural lands have also been burned and these lands have experienced increased overgrazing, salinization, and erosion. Continued destruction of forests, pastures, and agricultural lands will perpetuate soil degradation in Afghanistan. Wildlife populations continue to diminish. Water and air pollution are increasing. These environmental trends contribute to a declining quality of life for Afghanistan's population.

6

Historical Geography

James B. Dalton

Key Points

- Cultural diversity can be ascribed in part to the pre-eighteenth century movement of armies across Afghanistan.
- The Arab conquest in the seventh century produced a legacy of a shared faith with countries of North Africa and West Asia.
- The colonial period had a lasting impact on the country's borders and Afghani attitudes toward foreigners.

Historical geography is the study of the human geography of the past, and this chapter specifically examines Afghanistan's history as it relates to people who have controlled the space now occupied by the state of Afghanistan. The history of this region provides valuable insights into the present conditions and the current geopolitical reality, and we begin by reviewing the historical events before the current era. These early events helped to shape the environment for major actions in the eighteenth and nineteenth centuries, which subsequently led to the establishment of the state of Afghanistan. The interactions of colonial powers, Russia and Britain, are examined, as those confrontations led to conflicts within the region. The influence of colonial powers continued throughout the twentieth century, and the events of that period provide the context for understanding the current political geography of Afghanistan and the region (chapter 8).

Afghanistan's close proximity to "greater India" (India prior to partition in 1947) has contributed to its closely intertwined history. Over the past three thousand years, different Indian emperors have ruled various parts of Afghanistan. The physical geography (chapters 2–5) describes this land as the gateway to and from India. Consequently, Indian armies have conquered this area, and armies from outside the region have traversed this land en route to India. The list of peoples traveling through the area span a wide range: over the centuries, Alexander the Great, the Scythians, the White Huns, and the Turks all followed

31

the natural routes from central and west Asia through the region into India. When not being ruled or conquered by peoples from outside the region, different Indian rulers controlled portions of present-day Afghanistan. This country was also an important center of early Buddhism (Jackson et al., 2001).

THE PRE-EIGHTEENTH CENTURY ERA

The first major incursion into the region during the current era was conducted by the Arabs under the leadership of the Syrian Umayyads, who reached the area around 642 C.E. and took Kabul in 664 C.E. (Polk, 1991). This invasion and occupation introduced Islam to the region, and Islam eventually became the predominant religion. The Persians (Iranians) replaced Arab rule, around 750 C.E., and the Persians were subsequently replaced by the Turkic Ghaznavids in 998 C.E. The conquest, led by Mahmud of Ghazni, consolidated the Afghan Kingdom and established a great cultural center and a base for operations into India in search of riches. His rule and those of various princes after his death lasted until the Mongol invasions of 1219 C.E. After Genghis Khan's death in 1227, a succession of chieftains and princes controlled parts of present-day Afghanistan until the area fell under the relatively peaceful and prosperous rule of the Uzbek ruler Timur (Tamerlane) in the late fourteenth century. Under his successors, the Timurids, Afghanistan witnessed a transition period during the fifteenth century. In the early fifteenth century, Babur rose to power in Kabul and forcefully expanded to the east, establishing the Mughal Empire in greater India, which lasted in some form well into the nineteenth century. On the edge of the Mughal Empire, Afghanistan found itself caught, and sometimes divided, between the Mughals and the Safavid Empire in Persia (Iran).

THE EIGHTEENTH AND NINETEENTH CENTURIES

In the eighteenth century, the passing of the Moghul dynasty was followed by the rule of Ahmad Shah Durrani who was able to consolidate the fragmented elements of the region into what would become modern Afghanistan. He was from the Pashtun tribal confederation and first in a long line of Afghan rulers from this tribal federation that lasted until the Marxist coup d'état in 1978 (Jackson et al., 2001). The ability of one tribal group to dominate other tribal groups enabled the federation to exist. This form of organization depended on "highly personalized and charismatic power, and on precarious, fleeting agreements between tribes and tribal subgroups" (Tarock, 1999). The events of the eighteenth century established the conditions that the colonial powers of England and Russia would encounter as their respective empires expanded toward each other in the nineteenth century. Those conditions included tribes of people from Mongol, Persian, and Turkic ethnic backgrounds who shared a common religion and a closely related language, and were conditioned to living in a harsh environment.

By the nineteenth century, the British Empire was expanding out of India toward the north and west looking to offset what they saw as the expansion of Russia into Central Asia. Russia was also concerned with British expansion into areas that they saw as their sphere of influence. The British would be involved in three Anglo-Afghan wars, the first from 1832 until 1842 and the second from 1878 until 1880. These two wars were extremely ferocious, with the Afghan Afridi (a Pathan tribe) earning a reputation as "among the bravest, and certainly the cruelest, of the notoriously cruel Pathan Tribes" (Margolis, 2000). At the conclusion of the second Anglo-Afghan War, Britain and Russia agreed to the establishment of borders that would become the present-day Afghanistan, and Britain retained effective control over Kabul's foreign policy. The latter served as a catalyst that would spark the third Anglo-Afghan War (Jackson et al., 2001).

THE TWENTIETH CENTURY

From the end of the second Anglo-Afghan War until the end of World War I, Afghanistan remained neutral despite German encouragement. After the war, anti-British sentiment and the British control over Kabul's foreign policy led to a short third war (1919) with an attack into India. The British quickly gave up control of Afghanistan's foreign policy, ending the war in the same year it started. From 1919 through 1933, there were a series of assassinations and an abdication caused by attempts at reform and modernization, which alienated some conservative members of the clergy. In 1933, King Zahir Shah succeeded to the throne and reigned until 1973. He introduced a democratic constitution that allowed the formation of political parties. One of these parties was the People's Democratic Party of Afghanistan (PDPA) (Jackson et al., 2001).

At the end of his reign, King Zahir Shah was accused of corruption and blamed for the poor economic conditions. A military coup led to his demise in the summer of 1973, and a former prime minister declared himself president and abolished the monarchy. Despite numerous attempts at economic and social reforms, another bloody coup followed in 1978. It was this coup that brought the PDPA to power with the support of the Soviet Union. This government was fractured from within and was greatly resisted by a large number of Afghanis. It became apparent that the PDPA would be unable to survive without increased assistance from the Soviet Union. The assistance came in the form of ground troops in December 1979 that in essence replaced the PDPA leadership with those more willing to listen to the experts from the Soviet Union. This began the Soviet Union's long bloody conflict that would not end until 1989 (Jackson et al., 2001; Tarock, 1999). In the face of a common enemy, seven different groups in Afghanistan formed a loose alliance to fight against the PDPA and Soviet forces. When peace negotiations were eventually

Source: Eugene Palka

A destroyed Soviet tank, one of several hundred military vehicles and pieces of equipment that are strewn about the landscape in the Panjsher Valley, serves as a vivid reminder of the country's turbulent history.

conducted, these groups were left out. As a result, the fighting did not end with the Soviet Union's withdrawal in February 1989. The Soviet exodus removed the common enemy (the country's key centripetal or unifying force), thus returning Afghanistan to old tribal and factional rivalries (Jackson et al., 2001; Tarock, 1999).

THE POST-SOVIET AND TALIBAN ERA

The weak central government, installed by the Soviets, held on until 1992 when it was finally unseated in Kabul by the Mujahedin. Fierce fighting ensued between factions, with devastating consequences for the Afghan people. The widespread confrontations between rival tribes during 1993 left more than 10,000 civilians dead and Kabul was reduced to rubble (Rashid, 2001).

Despite numerous diplomatic missions in 1994, the fighting continued and hopes of a peaceful resolution appeared to be beyond the realm of the possible. An incident that changed the course of history occurred in November 1994, when a Pakistani truck convoy en route to Turkmenistan was held by warlords in Kandahar. It had been common practice for warlords to demand tolls for passage along this and other routes. Between Chaman and Herat alone, there were 71 "check points" where money was extorted for promises of "safe passage" (Matinuddin, 1999). A relatively unknown group referred to as the Taliban attacked the warlord and his band of thugs, freed the convoy, and took

Historical Geography

Source: Cecilio M. Ricardo Jr.

The nationwide celebration of Eid has begun, and girls have dressed up for the occasion, December 17, 2001. During the Taliban rule Afghanis were not allowed to celebrate Eid or other holidays, nor were girls permitted to dress up in colorful clothes, wear earrings and lipstick, and carry accessories such as purses.

control of Kandahar in a four-day battle (Rashid, 2001). The Taliban were a group formed and led by a charismatic individual, Mullah Mohammad Omar, who was guided by his strict interpretation of Islamic law. The Taliban emerged as a dominant political and military force in Afghanistan by the end of 1994. Over the course of the next four years, they expanded their control outward from Kandahar and the southern provinces. Operating out of the south, they extended control to the west, seizing Herat in 1995, and then north toward Kabul in 1996. Kabul eventually fell to Taliban forces on 26 September 1996 after a series of attacks and counterattacks by the opposition (Rashid, 2001).

With control of Kabul, the Taliban came to be regarded as a stabilizing force from some perspectives, while others viewed the regime as just another

Source: Andres Rodriguez

A local Tajik is one of many who has been hired to apply his skills as a carpenter to help rebuild the airbase at Bagram.

despotic warlord-led faction (Rashid, 2001). At the time, the Taliban ushered in a brief period of stability for a large proportion of the population, but it came at a high cost in terms of human rights and individual freedoms. Women suffered the greatest injustices as a result of the strict Taliban rule. But despite their barbaric methods of enforcement, the Taliban never succeeded in gaining uniform control over the country. Moreover, the Taliban failed to claim international recognition as the legitimate government of Afghanistan because of their widespread abuse of human rights, religious persecution, involvement with opium production, and support of terrorist organizations, specifically al-Qaeda. As the Taliban tightened their grip, even countries like Pakistan, an early supporter of the regime, started to distance themselves from the Taliban government (Rashid, 2001).

Sporadic fighting continued in the north throughout the late 1990s and into 2001, fueled by opposition forces. The opposition was composed of a

Historical Geography

Source: Eugene Palka

The airfield at Bagram has been a battleground between Soviets and Mujahedin, Northern Alliance and Taliban, and United States and al-Qaeda. The collection of destroyed aircraft provides clear evidence of the area's turbulent history.

shifting coalition made up of ex-communist, regional warlords, and other minority ethnic tribes unwilling to join the strict Islamic movement of the Taliban. The coalition became known as the Northern Alliance, in part due to their geographic position in the northeastern part of the country. The Taliban era came to an end in part because the opposition ultimately joined the United States and Coalition forces, who eventually unseated the Taliban in late 2001 and early 2002.

SUMMARY

This brief discussion helps to illuminate four key areas of the present situation in Afghanistan. First, the cultural diversity (see chapter 7) can be explained in part by the pre-eighteenth century movements of armies to and from India, imparting language, religion, and political influence. Second, the current ties with Pakistan are as much a result of the historical periods where Afghanistan was a part of and/or controlled by greater India, than as a result of current geopolitical concerns of Pakistan. Third, the introduction and adoption of the Islamic faith during the Arab conquest produced a legacy of a shared faith with countries of North Africa and Western Asia. Lastly, the colonial period had a lasting impact on Afghanistan, specifically reinforcing the latter's desire to remain outside of the dominance of colonial powers. It was the fierce determination of the Afghan tribes to remain free from the British that established the strong tra-

Historical Geography

Source: Cherie A. Thurlby

The traditional *burqa* remains popular with women, especially in rural areas.

dition of resistance to foreign rule. This tradition continued through the Soviet occupation. Most recently, the Taliban phenomenon changed the internal politics and way of life in profound ways and served to stir international concern. Since the fall of the regime another era of intervention has ensued and a new form of government is in place. This volatile country remains divided, however, perhaps until a common enemy threatens the political entity and promotes a loose sense of unity among its diverse peoples.

7

Cultural Geography

Jon C. Malinowski

Key Points

- With over 99 percent of the population Muslim, Afghanistan's religious landscape includes areas of strict fundamentalism.
- Afghanistan's rugged terrain and peripheral location enabled ethnic groups to remain relatively isolated throughout history, resulting in considerable ethnolinguistic diversity.
- Pashtuns and Tajiks are the two largest ethnic groups, totaling over 60 percent of the population.

Afghanistan's cultural landscape is a mottled pattern painted by history and geography. Wedged between major culture hearths and repeatedly crossed over by invaders, the area presents a cultural landscape that is as dramatic and varied as its forbidding terrain. Cultural geographers study both the patterns of cultural traits, such as religion and language, and also the way culture modifies the natural and built environment. This chapter addresses the distribution of major cultural patterns such as ethnolinguistic groupings and religion, but it must be remembered that these basic traits affect much more than communication and spiritual beliefs. Religion alone affects dietary habits, clothing, architecture, holidays, the workweek, schooling, and criminal justice, among other aspects of daily life. So, an understanding of Afghanistan's basic cultural patterns is the first step toward an awareness and appreciation of the very fabric of Afghani lives, landscapes, and longings.

RELIGION

It is commonly known that Afghanistan is primarily a Muslim country. In fact, followers of Muhammad's teachings make up over 99 percent of the population (Britannica.com, 2001). Islam likely entered Afghanistan over 1,200 years ago, but the current religious landscape has been affected by many events over

Cultural Geography

Source: Eugene Palka

Children at play outside the village of Charikar

the centuries. Invaders brought new forms of religion, and the rise and fall of dynasties changed the nature of spiritual life in the region. The destruction of 2000-year-old Buddhist statues by the Taliban government in early 2001 reminds us that early religious practices flourished before Islam, and the same is true of modern Afghanistan as evidenced by isolated Hindu and Jewish communities in the country today (*Economist*, 2001). The presence of a small number of Sikhs, followers of a religion that arose in India in the sixteenth and seventeenth centuries, shows that later ideas and belief systems also penetrated the mountainous country. But by and large, Afghanistan is a Muslim country.

The vast majority of Muslims in Afghanistan are followers of the Sunni branch of Islam, which accounts for about 90 percent of all Muslims worldwide (see Figure 7.1). Unlike most Sunnis in other Muslim countries, some of Afghanistan's Sunni Muslims, including members of the Taliban, are quite fundamental and conservative in their practice. The fundamentalist ideas of Wahhabism from Saudi Arabia and Deobandism from Pakistan have become increasingly important among some Afghani clerics and rulers (Kaplan, 2000). Wahhabism has its origins in the eighteenth-century teachings of the Saudi cleric Muhammad ibn 'Abd al-Wahhab and is the officially accepted form of Islam in Saudi Arabia. Deobandism arose during British rule in South Asia as a reaction to colonial ideas. Refugees fleeing to Pakistan during the Soviet invasion of the 1980s were often taught these more extremist views of Islam in refugee camp schools and colleges (Kaplan, 2000).

Cultural Geography

Figure 7.1 Islam in Afghanistan and Surrounding Countries

Source: CIA, 1995

The balance of Muslims in Afghanistan comes from two sects of the other major branch of the religion, the Shi'ites (see Figure 7.1). The Shi'ite/Sunnite split in Islam was the result of political differences within the religion in the years following the death of Muhammad in 632 C.E. Shi'ites were followers of Ali, Muhammad's son-in-law, and after his death argued that the leader of the religion, the *caliph*, should come from his lineage. Sunnites felt more pragmatically and accepted caliphs drawn from leaders within the community (Awn, 1984).

Most of the Shi'ites in the country are from the sect commonly referred to as *Twelvers*. Within the early Shi'ite community, the heir to Muhammad's authority was known as the *imam* (not to be confused with the modern usage of this word by Sunnites to refer to the leaders of prayer at a mosque). Twelvers accept a line of twelve imams, beginning with Ali, and ending with the disappearance of a child around 874 C.E. (Awn, 1984). Some Shi'ites, however, disputed the authority of the seventh imam, Musa, during the eighth century. This faction backed the legitimacy of Musa's brother Isma'il. Today, followers of this splinter group are known as *Isma'ilis* or *Seveners*. While a discussion of the particulars of early Muslim history may seem academic, it reminds us that all Afghani Muslims should not be considered to have the same beliefs, practices, or attitudes.

Most Muslims do, however, share a core of beliefs based on the five "pillars" of Islam. The first pillar is a profession of faith that Allah is God and Muhammad His prophet ("There is no god but God, Muhammad is the messenger of God"). Through Muhammad, Allah gave the world his word, recorded word-

Cultural Geography

Source: Preston E. Cheeks

Pilgrims from all over Afghanistan board Ariana Airlines to fly to Mecca during a Hajj. The pilgrims arrive in Kandahar and fly from Kandahar Army Airfield to Saudi Arabia, where they will participate in various religious rituals to fulfill one of the five pillars of Islam.

for-word in Arabic on the pages of the Qur'an (Koran). For this reason, Muslims worldwide study Arabic in order to understand the teachings of God in its original, and thus pure, form.

The remaining pillars of Islam outline basic practices of daily and religious life. A Muslim must pray five times a day in the direction of the Ka'ba, the sacred cube known as the Holy House of God in Mecca, Saudi Arabia. Mecca is the birthplace of Muhammad and the holiest city in the faith. Prayer, preceded by ritual washing, takes place at daybreak, noon, midafternoon, sunset, and during the evening. Prayer does not have to take place at a mosque, but many Muslims do attend a prayer service at a mosque at noon on Friday. Because of this practice, many banks, stores, and government offices will close on Friday. The third pillar requires a Muslim to give alms to the religious community and to the needy; ten percent of one's annual income is a number sometimes quoted, but the actual number varies greatly. The fourth pillar mandates fasting during the daylight hours of the 28-day month of *Ramadan*, which changes yearly because it is based on a lunar calendar, but generally falls during the summer. The fifth and final pillar mandates that every Muslim who is able must make a pilgrimage, the *Hajj*, once during his or her life to the holy city of Mecca in Saudi Arabia. There, with Muslims from all over the world, adherents complete a multiday sequence of physical and spiritual feats.

Another aspect of Islam that should be understood is the importance of Islamic law, or *shari'a*. Drawn from the Qur'an, the customs of Muhammad

Source: John Wiegand

Old habits die hard, as evidenced by the women who continue to wear the traditional *burqa* even after the Taliban were unseated.

(*sunna*), tradition (*hadith*), academic and theological reasoning, and community consensus, shari'a rules govern both religious life and the common, everyday lives of Muslims (Mayer, 1984). In largely secular Muslim countries, such as Indonesia or Turkey, shari'a has been replaced by more universal legal systems, but under Taliban rule in Afghanistan it was the primary legal structure. Conservative forms of Islamic law dictate dress codes, such as the wearing of veils for women (the *burqa*) and beards for men. Alcohol is prohibited, as is the eating of pork. All meat must be slaughtered in a specific way; properly prepared food is known as *halal.*

Under Taliban rule, Islamic law was strictly enforced and, some would argue, distorted to a barbaric degree. Stories of executions and other forms of corporal punishment have been widely reported in the media. As an example, thieves commonly have a hand lopped off for their crime. Kite flying, Western music, and television were illegal under Taliban control. People were even beaten or chased with dogs for improper dress or a beard that was too short (Shah, 2001). Homosexuality was regarded as a crime, and like other crimes, might be dealt with by throwing the accused off a high building or having a wall collapsed on them (Shah, 2001). Converts to other religions were sometimes executed.

Islam also affects the built environment. Mosques are common in almost every neighborhood. Ranging from elaborate to simple, a common feature is the ubiquitous *minaret*, or tower, that enables spiritual leaders to call worshippers to prayer. Mosques serve not only as a place of religious gathering, but also as a community center, personal retreat, and after-school center. Many large towns have bigger, more elaborate mosques at the center of the city. *Madrasas*, Islamic schools or colleges that have been educating young Muslims for centuries, are often located near the larger mosques.

ETHNOLINGUISTIC GROUPS

There is no single Afghani people. The country's location between the former empire states of Central Asia, Persia (Iran), and South Asia ensured that invading groups settled in the area as they moved through or temporarily controlled territory. In addition, the rugged terrain of the country enabled groups to remain isolated from each other, thus slowing the process of acculturation and the sharing of cultural traits (Allan, 2001). *Ethnologue*, a catalog of world languages, identifies 45 languages within the country (Ethnologue, 2001). Because certain languages or families of languages are generally associated with particular ethnic groups, this discussion will look at the most important *ethnolinguistic* groups of Afghanistan. As an aside, it is worth noting that none of Afghanistan's ethnic groups are Arab, and that outside of religious services, the Arabic language is spoken by only a few thousand people.

The largest ethnolinguistic group in the country is the Pashtuns, also known as Pushtuns, Pakhtuns, or Pathans, who make up 38 percent of the population (Central Intelligence Agency, 2001). Before history created a country of Afghanistan, the Pashtuns were known as Afghanis. They speak an Indo-European language called Pashto, one of two official languages in the country, spoken by about 35 percent of all citizens (CIA, 2001). While there are nearly 8 million Pashtuns in eastern and southern Afghanistan (Figure 7.2), there are twice as many in neighboring Pakistan, linking these two countries (Figure 7.3). The Taliban were predominantly Pashtun, which explains, in part, the support of their rule from many Pakistanis across the border.

Within Pashtun areas, communities are organized into clans known as *khels*, which are then organized into tribes known as *kaum* or *qabili* (Lieberman, 1980). Scores of different tribes occupied separate areas, and historically tribal conflicts have been common. Tribes are also linked together through confederations that are sometimes at odds with each other. Most notably is the periodic struggle between the large Durrani confederation of mostly sedentary tribes in the south and the more nomadic Ghilzais group in the east (Lieberman, 1980). In terms of religion, most Pashtuns are Sunni Muslims, but it should not be assumed that all Pashtuns are as fundamentalist as the Taliban in their views of Islam. There is also an ethical code among some Pashtuns, especially in rural areas, known as Pashtoonwai, or the "way of the Pashtuns." This code of conduct both encourages hospitality and chivalry while also calling for vengeance or strict penalties for crimes committed against a fellow member of a clan or tribe (Kaplan, 2000).

Figure 7.2 Ethnolinguistic Groups in Afghanistan

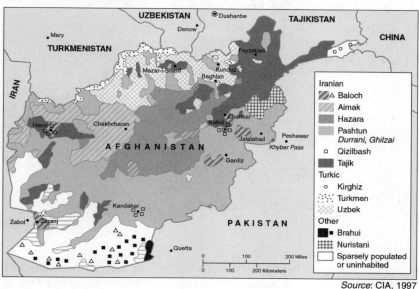

Source: CIA, 1997

The second largest ethnic group in the country is the Tajiks, accounting for about 25 percent of the population (CIA, 2001). Like the Pashtuns, they speak an Indo-European language, Dari, an eastern dialect of Farsi (Persian or Iranian). Overall, about half of all Afghans speak Dari according to the Central Intelligence Agency (2001). Tajiks have millions of ethnic brethren in neighboring Tajikistan, Uzbekistan, and China. Within Afghanistan, Tajiks are concentrated in the northeastern part of the country with pockets in the west (see Figure 7.2). They formed an important part of the groups that challenged Taliban rule in 2001 and 2002. Although Tajiks are Sunni Muslims, they generally are considered to be less fundamentalist and conservative than the former Taliban government.

A third, important group in the country is the Hazara, who inhabit much of the central part of the country (Figure 7.2). Numbering about 3 million and accounting for just fewer than 10 percent of the population (CIA, 2002), Hazaras speak Dari like the Tajiks and other groups. Many Hazara are Shi'ite Muslims, which, along with other cultural differences, put them at odds with the Taliban. The Taliban's persecution of Hazara within Afghanistan caused a great deal of concern within Iran. A 1998 massacre in the city of Mazar-i-Sharif of hundreds of Shi'ites pushed Iran and Afghanistan to the brink of war (Tarock 1999).

Major remaining ethnic groups in Afghanistan include Nuristani, Turkmen, Uzbek, Kirghiz, Chahar Aimak, Brahui, and Baloch groups. The Nuristani inhabit a region of eastern Afghanistan and western Pakistan known as Nuristan. Formerly known as the Kafir (the Arabic word for infidel), this

Cultural Geography

Figure 7.3 Ethnolinguistic Groups in Pakistan and Surrounding Areas

Source: CIA, 1980

group was forced to convert to Islam during the late nineteenth century and given the name *Nuristani*, meaning "enlightened" (Britannica, 2001). Due to their isolation and historical differences, Nuristanis remain culturally different from the rest of Afghanistan, and despite their conversion to Islam, polytheism and animal sacrifice are still practiced. Like others in the region, their language, Nuristani or Kafiri, is also an Indo-European language.

The Turkmen, Uzbek, and Kirghiz (Kyrgyz) groups share cultural areas with majority groups in neighboring Turkmenistan, Uzbekistan, and China. In general, they are later settlers than other groups in the country and possess distinct cultural differences. As an example, they all speak Turkic languages that are different from the Indo-European tongues spoken by Pashtuns, Tajiks, Hazara, and the like. Islam is a unifying feature to some degree, but these Central Asian minorities often have considerable disagreements with Pashtun fundamentalism. Uzbeks alone account for about 6 percent of the population (CIA, 2001).

The Aimak of western Afghanistan are notable because they probably are of Mongolian origin, a reminder of past invasions. They have Mongoloid features and live in yurts similar to those found among groups in Mongolia and western China (Britannica, 2001). Despite their eastern Asian origins, however, they speak the Farsi dialect of Dari.

Finally, the Baloch (Baluchi) and Brahui groups live in a large area that includes parts of southern Afghanistan, Iran, and Pakistan (see Figures 7.2 and 7.3). Known as Balochistan (or Baluchistan), this dry, isolated region prompts many to live as nomadic herders of sheep. This isolation has also spawned a distrust of central, outside authority, and the Baloch have periodically made attempts to gain greater autonomy from the governments of Pakistan, Afghanistan, and Iran. Most Baloch are Sunni Muslims and speak an Indo-European

Cultural Geography

Source: Andres Rodriguez

Local residents have flocked to U.S. military bases in search of jobs. These Tajik men have found work at Bagram airbase, north of Kabul.

language that is somewhat related to Kurdish. They number about 250,000 in Afghanistan alone and nearly 8 million across the region. The Brahui are a smaller group of about 2 million in all countries who live among the Baloch (Ethnologue, 2000). They speak a language that is Dravidian in origin, a language family that dominates in southern India and which includes Tamil and Telugu. Because of its isolation from languages in the same family, and owing to the fact that it is not a written language, Brahuis have borrowed many words from other languages and bilingualism is common (Ethnologue, 2000).

SUMMARY

Afghanistan is a diverse country. Although most are Muslims, different sects and degrees of fundamentalism shatter any notion of religious uniformity. Although ethnic Pashtuns have generally dominated the country, they remain less than 40 percent of the population. Two languages are each spoken by over 35 percent of the country and several major world language families are represented. Ethnicity ties Afghanistan to the western states of Iran and the Middle East, to the newly independent Central Asian republics, and to the demographic powerhouses of Pakistan and India. Afghanistan is truly wedged between cultural boundaries, creating puzzle-like cultural patterns that defy simple definitions.

8

Political Geography

Andrew D. Lohman

Key Points

- Afghanistan has a long history of political instability.
- Its prolonged history of conflict has led some to characterize Afghanistan as ungovernable.
- After leading the country's interim government following the fall of the Taliban regime, Hamid Karzai was elected president in June 2002.

Political geography is the spatial analysis of political phenomena and processes. Essentially, this field studies the interaction of political processes across geographic areas, which may occur at every scale of analysis, from the local to the global level. Within political geography, the most basic and common unit of analysis is the state, which is the dominant form of political organization in the world (Glassner, 1996). This chapter will discuss Afghanistan's internal political arrangements as well as its geopolitical position in the Central Asia region.

AFGHANISTAN: THE STATE

As the fundamental unit of analysis in political geography, the state is defined as a politically organized territory administered by a sovereign government and recognized by a significant portion of the international community (de Blij & Muller, 2001). Originally created to serve as a buffer zone between Russian and British colonial spheres in Central Asia, Afghanistan's current boundaries were delimited when the Afghan rulers signed agreements with the Russian government in 1885, defining the Oxus River (now known as the Amu Darya) as Afghanistan's northern boundary. Subsequent agreements were made with the British Raj in India in 1893 (Margolis, 2000).

With over 5,529 km/3,317 miles of land boundaries, Afghanistan shares borders with six other states (China 76 km/47.2 mi, Iran 936 km/581 mi, Pakistan 2,430 km/1,509 mi, Tajikistan 1,206 km/749 mi, Turkmenistan 744 km/462 mi, and Uzbekistan 137 km/85 mi) (CIA *World Factbook*). Although vir-

Political Geography

Figure 8.1 Political Divisions in Afghanistan

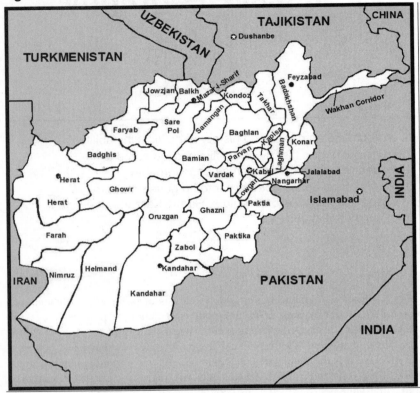

Source: Andrew D. Lohman

tually compact in shape, Afghanistan is actually a prorupt state, with an extension of territory known as the Vakhan Corridor that separates Pakistan from Tajikistan and provides Afghanistan with a narrow border with China in the rugged Hindu Kush Mountains (see Figure 8.1).

Although drawn by British and Russian representatives in the nineteenth century, the boundaries have proved fairly stable and undisputed. These boundaries, however, essentially divided several nations of people and have created a country out of a heterogeneous mixture of ethnolinguistic groups. This complex cultural geography, together with the difficult physical landscape has led Kenneth Weisbrode to describe Afghanistan's internal political situation as "ungovernable" (Weisbrode, 2001). Although Afghanistan has been embroiled in civil war since the late 1970s, a period of intense political turmoil followed the overthrow of the Soviet-sponsored government of Ahmedzai Najibullah.

Following the collapse of the Soviet Union and facing a multitude of internal challenges, Russia ended its financial, moral, and material support for the communist government in Kabul in January 1992. Without this foreign aid,

Political Geography

In the province of Parwan, people commemorate the life and death of former Northern Alliance leader, Ahmad Shah Masood by displaying posters, banners, and flags.

the Najibullah government could not withstand the pressure from the various ethnic rebel forces vying for control of the state, and in April 1992, Najibullah was forced to cede control of the country (Margolis, 2000). In the ensuing power vacuum, three of the main resistance factions united to form the Northern Alliance, which consisted of a Tajik faction led by Ahmed Shah Massoud, an Uzbek faction led by Abdul Rashid Dostam, and the Hazara group Hizb-i-Wahdat (Weisbrode, 2001). Together, these groups established a new government under the leadership of Buhanuddin Rabbani, but the Alliance excluded Ghilzai Pashtuns from the newly formed government, prompting renewed fighting between the various ethnic groups (Weisbrode, 2001).

In the ensuing civil war, each ethnic group struggled to consolidate and maintain control over its regional base areas with Dostam in Mazar-i-Sharif, Ismail Khan in Herat, and Massoud in the Panjshir Valley region (northeast of Kabul in the region where Parvan, Kapisa, and Laghman provinces meet, and extending northeast into Badakhshan) (Weisbrode, 2001). Meanwhile, the Taliban emerged onto the political scene from Kandahar in southern Afghanistan and from the refugee camps in Pakistan.

The Taliban (the plural form of *talib*, student) was an Islamic fundamentalist group, composed predominately of Pashtuns. Drawn from Islamic *madrasas* (religious schools) in Kandahar and the refugee camps along and inside the Pakistani border, the Taliban was formed with the expressed aim of ridding the country of lawlessness, banditry, and atheism (Margolis, 2000). Margolis

Table 8.1 The Provinces of Afghanistan

Province	Area (mi^2)	Provincial Capital
Badakhshan	18,298	Feyzabad
Badghis	8,437	Qal'eh-ye Now
Baghlan	6,604	Baghlan
Balkh	4,861	Sharif
Bamian	6,722	Bamian
Farah	18,446	Farah
Faryab	8,600	Meymaneh
Ghazni	9,024	Ghazni
Ghowr	14,925	Chakcharan
Helmand	23,866	Lashkar Gah
Herat	23,668	Herat
Jowzjan	7,326	Sherberghan
Kabul	1,770	Kabul
Kandahar	18,403	Qandahar
Kapisa	722	Raqi
Konar	4,045	Asadabad
Kunduz	3,021	Konduz
Laghman	2,783	Mehtarlam
Logar	1,796	Baraki
Nangarhar	2,940	Jalalabad
Nimruz	15,963	Zaranj
Paktia	3,698	Gardeyz
Paktika	7,464	Zareh Sharan
Parwan	3,628	Charikar
Samangan	5,969	Samangan
Sare Pol	9,856	Sare -Pol
Takhar	4,777	Taloqan
Uruzgan	11,308	Tarin Kowt
Wardak	3,483	Kowt-e Ashrow
Zabul	6,675	Qalat

Source: World Geographical Encyclopedia, vol. III, Asia, 1995

(2000) further contends that the Taliban was initially the product of Pakistan's Inter-Service Intelligence, created to rid Afghanistan of anarchy and banditry, but also to prevent such behavior from spilling over the border into Pakistan. Weisbrode (2001) doesn't specifically state that Pakistan created the Taliban, but he does describe how the Pakistani government supported the Taliban and intended to use the fundamentalists to help Pakistan open a road from Islamabad through Herat and into Turkmenistan, thus opening a highway for trade between Pakistan and Central Asia. Along with this highway, the Pakistanis also had visions of a pipeline from Central Asia through Pakistan to the Arabian Sea, but such ventures would only occur if the violence and civil war in Afghanistan ended (Margolis, 2000).

By March 1995, the Taliban had effectively seized control of Kandahar and the twelve surrounding provinces, and these successes were in large part due to Pakistani assistance (Margolis, 2000). Armed with such support, the

Taliban expanded their control to the point where various government and news agencies estimated that the Taliban controlled approximately 90 percent of the country prior to U.S. intervention in 2001.

Meanwhile, the United States and the United Nations continued to recognize Buhanuddin Rabbani as the legitimate leader and government of Afghanistan during the Taliban era. Under Rabbani's leadership, the country's official name was the Islamic State of Afghanistan, was ruled by an Islamic Council, and the government comprised elements of the Northern Alliance.

During their ascent to power, the Taliban effectively enforced a strict adherence to their narrow interpretation of the tenets of the Koran and advocated a return to the traditional ways of the Islamic faith. Intent on creating a true Islamic society, the Taliban renamed the country as the Islamic Emirate of Afghanistan (Stump, 2000). Pakistan was the only country that recognized the Taliban regime as the legitimate government in Afghanistan (Saudi Arabia and the United Arab Emirates had also recognized the Taliban leadership in Afghanistan but revoked recognition in the wake of the terrorist attacks in the United States).

Led by Mullah Mohammad Omar, the Taliban indirectly ruled the country, which is divided into thirty provinces (see Figures 8.1 and Table 8.1), through local councils that administered the traditional Islamic law, or *sharia*. Under the sharia, traditional punishments for crimes include amputations for theft, stoning for adultery, and execution for a number of other crimes. To enforce the laws, the Taliban created a form of religious police (known as the Department for the Propagation of Virtue and the Prohibition of Vice), which was also charged with enforcing restrictions on Western influences, such as dancing, movies, television, and popular music, within the country (Stump, 2000).

In striving to create their version of a pure Islamic society, the Taliban were reported to be ruthless in their enforcement of their ideals. Margolis (2000) considered the Taliban's brutality to be a reaction to the Communist era. Under the Soviet/Russian-sponsored communist regime of the 1980s and early 1990s, social modernization, education, and equality for women were imperatives for creating a true communist state in Afghanistan, yet many Afghanis considered each of these initiatives to be violations of the tenets of Islam. In order to prevent these forces from irreparably changing the nature of Afghan society, the Taliban used every means necessary to ensure that the people follow their strict interpretation of Islamic law. Not all Afghans, however, embraced the beliefs and ideals of the Taliban. Although the Taliban initially found support in their drive to create a true Islamic society, mostly from the Pashtun sectors of the population, Stump (2000) described how the Taliban's strict enforcement of traditional beliefs "alienated Afghans who didn't follow the strict interpretation of Sunni Islam."

A U.S-led coalition successfully ousted the Taliban from power in November 2001. A month later, a number of prominent Afghans met under UN auspices in Bonn, Germany, to decide on a plan for governing the country. As a result, the Afghan Interim Authority (AIA), made up of 30 members and headed by a chairman, was inaugurated on 22 December 2001 with a six-month mandate followed by a two-year Transitional Authority (TA), after which elections are to be held (CIA, 2002). The structure of the follow-on TA was announced on 10 June 2002, when the Loya Jirga (grand assembly) convened establishing the Transitional Islamic State of Afghanistan (TISA), which has an 18-month mandate to hold a Loya Jirga to adopt a constitution and a 24-month mandate to hold nationwide elections (CIA, 2002). Since 10 June 2002, Hamid Karzai has served as the president and head of government.

A GEOPOLITICAL ANALYSIS

Afghanistan's current political situation is fragile while in transition. Aside from the chaotic political legacy he has inherited, President Karzai must contend with the immediate aftermath left behind by the Taliban regime. Apart from imposing their own ideals and interpretation of the Koran on the Afghan people, the Taliban at various times, "espoused expansionist philosophy," and their policies and beliefs "antagonized neighboring countries" (Stump, 2000). Although, the Taliban have been removed from power, and replaced by an entirely new regime, neighboring countries continue to fear that other extremist groups may wish to push their fundamentalist agenda at a later date, if they are ever able to regain control of the country.

Although landlocked and seemingly remote and isolated, Afghanistan has always occupied a position of geopolitical significance in Central Asia, since the country lies at the crossroads of three great geopolitical regions, the Indian subcontinent, Iran, and Central Asia (Parker, 1995). Weisbrode (2001) argues that "nearly every transitional problem in Central Eurasia is linked to the Afghan civil war." Although each neighboring country has generally tried to stay out of the turmoil within Afghanistan, in many cases neighboring states have felt obligated to support or assist one side or another in order to protect their own internal stability.

Pakistan is known to have assisted the Taliban in the early stages of their development in order to help control the lawlessness within the country and to prevent political dissent from spreading into Pakistan. Meanwhile, Russia, India, China, and the former Soviet republics of Turkmenistan, Tajikistan, and Uzbekistan have supported or assisted one or all of the many factions of the Northern Alliance in order to contain the fighting and prevent the Taliban from exporting their religious fundamentalism.

Although Turkmenistan, Tajikistan, and Uzbekistan are now independent states, these countries are still reliant on Moscow to a certain degree, based on previous economic ties and cooperatives established under the Soviet system. Moreover, Russia still considers its "strategic borders" to coincide with the boundaries of the former Soviet Union (Margolis, 2000). And, with these strategic borders adjacent to Afghanistan, Russia fears the expansion of the Islamic fundamentalist ideals, which if they take root in the former Soviet repub-

Political Geography

Much of the rural landscape reflects a history of warfare. The area pictured here was a battleground between the Afghanis and Soviets, the Northern Alliance and the Taliban, and the United States and al-Qaeda.

lics, could threaten Russia's relations with the former republics and jeopardize Russia's position in Central Asia. The Taliban was also believed to harbor, train, and support Chechen fighters in their conflict against Russia.

India and China also supported the forces arrayed against the Taliban in order to protect their own interests. In addition to the geopolitical stances and alliances, China and India have aided the forces of the Northern Alliance because the Taliban allegedly supplied fighters, assistance, and funding to rebels and terrorists of Xinjiang province in China, the Philippines, Yemen, and Indonesia (Weisbrode 2001). Additionally, the Taliban has been accused of providing a great number of fighters and assistance to the Muslim insurgents in Kashmir, who are fighting against Indian control of that region.

Iran assisted the Northern Alliance forces as the latter fought to resist Taliban advances. Besides the conflict between Shiite Iran and the Taliban's fundamentalist Sunni Muslim ideals, Iran hopes that any natural resource pipelines from Central Asia will travel through Iran (and not through Pakistan) to ocean ports and the world market.

SUMMARY

Politically, Afghanistan has long encountered a significant number of internal and external challenges to effective government and stability. The country has been wracked by civil war that postured ethnic groups against each other. Previously, it had no constitution or operating legislative body, and the Islamic courts under the *shari'a* were the expressed governing principles, although the latter are not fully embraced by all sectors of society (CIA World Factbook). Any attempts at resolving the conflict and achieving a peaceful, cooperative diplomatic solution between the various factions proved futile. Now in the aftermath of the Taliban regime, President Hamid Karzai is confronted with an enormous range of challenges as he attempts to exercise uniform control over the country and to lead a turbulent, war-torn, and poverty-stricken Afghanistan into the twenty-first century.

In addition to hosting the longest running civil war in the world today, Afghanistan is currently the society with the greatest number of refugees (over 5 million), the leading producer of opium in the world, and it has the greatest number of land mines within its borders (estimated at over 10 million, from the Soviet occupation as well as the civil war that ensued) (Weisbrode, 2001). Singly, each of these issues presents a number of obstacles to political, social, and economic development in Afghanistan in the near future. Collectively, they paint a dismal picture for the country's future prospects.

Source: Vernell Hall

A girl who was left homeless due to torrential flooding stands along the road. Villages in the Parwan Province were flooded by the Panshir and Ghorband Rivers. Personnel from the U.S. Civil Affairs Battalions provided humanitarian assistance to her and others in the province in April 2003.

9

Economic Geography

Albert A. Lahood

Key Points

- Years of international economic sanctions and internal mismanagement have resulted in subsistence-based and black market export economies.
- Although opium is the primary export, the majority of poppies are cultivated by independent farmers as a subsistence cash crop.
- Afghanistan has sufficient fossil fuel reserves, but there is insufficient infrastructure to exploit this potential component of their economic base.

Afghanistan's economic situation is reflective of years of inter- and intra-state conflict, and international isolation of the Taliban government. The result has been an economy divided into two spheres, an international illicit narcotics production and export economy, and a local subsistence agricultural economy, supplemented by local and regional periodic markets. Although there are sufficient fossil fuels available, the instability and paucity of infrastructure have limited the state's ability to exploit fossil fuels, resulting in the underdevelopment of this potential source of revenue. Internal mismanagement, international interventions, and isolation have prevented Afghanistan from participating in the legal international economy. In the aftermath of the Taliban era, change will be slow, if it does occur in these areas. The current government of Hamid Karzai has inherited a legacy of economic devastation and must draw from international resources and nongovernment agencies if he hopes to guide the country into a new period of relative economic self-sufficiency.

Economic geography is essentially about people and their struggle to make a living. This endeavor, however, is enhanced or constrained by ecological factors such as soil type, climate, vegetation and the availability of technologically exploitable natural resources. Additionally, structural factors such as transportation networks, financial institutions, governmental intervention, and interstate relations affect internal and external economic relations across space.

Economic Geography

Figure 9.1 Afghanistan's Poppy Cultivation

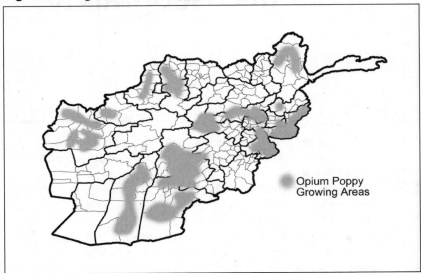

Source: Data from the United Nations, 2001

Economic geography can be thought of as the study of the various ways in which people earn a living or subsist, and how the goods and services they produce are organized spatially at different scales.

This chapter will first consider the current economic conditions of the Afghan state and its economic potential by reviewing the agricultural, industrial, and transportation sectors, as well as its primary resource base. Secondly, the current economic conditions will be explored in terms of how they affect the Afghani people and their everyday lives. Finally, there will be an overview of the development of illicit opium production and trade that is now the primary economic enterprise of many Afghan households.

AGRICULTURE

Today, agriculture is the largest segment of the Afghan economy. Estimates indicate that 67–85 percent of Afghanis are engaged in subsistence and commercial farming (*Afghanistan Online*, 2001; CIA, 2002; *CountryWatch*, 2002). The primary regions of agricultural production are the irrigated lands centered along the Kabul, Helmand, Harut, and Darya-ye Rivers, as evidenced by the increase in poppy production in the Kabul, Nangarhar, Helmand, Heart, and Badakhshan provinces (see Figures 9.1 and 9.2) (United Nations, 2001). Of Afghanistan's 647,500 sq km (249,935 sq mi), only 12 percent or 77,700 sq km (29,992 sq mi) are arable. This equates to a physiological density of roughly 250 people per sq km of arable land who are attempting to eek out a living. The

Economic Geography

Figure 9.2 Land-Use Patterns

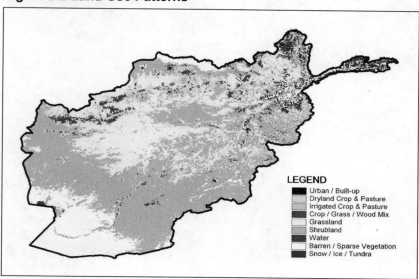

LEGEND
- Urban / Built-up
- Dryland Crop & Pasture
- Irrigated Crop & Pasture
- Crop / Grass / Wood Mix
- Grassland
- Shrubland
- Water
- Barren / Sparse Vegetation
- Snow / Ice / Tundra

Source: Brandon Herl

statistic does not, however, account for the pastoral alpine lands used for summer grazing of sheep and goats.

Although Afghanistan has a high physiologic density, the state has the capacity to meet its population's food requirements and has done so historically. The country, however, experienced a large decline in crop production between 1998 and 2000. This is attributed to a decrease in annual precipitation due to an abnormally strong Siberian high-pressure center that blocked moisture-laden westerly-tracking low-pressure systems (Palka, 2001). This phenomenon resulted in a 60 percent shortfall of Afghanistan's annual food requirements and prompted the United Nations to begin emergency food drops in June 2000 in an effort to mitigate the impact of the drought (*CountryWatch*, 2003, CIA, 2002). In 2001, however, a 10 percent increase in overall food crop production was recorded.

Sheep- and goat-herding is another important part of the Afghani agricultural economy, producing 356.8 metric tons of meat per annum. Until the recent increase in poppy cultivation, herding was the primary source of household income (*CountryWatch*, 2003). This portion of the Afghani's agricultural sector is concentrated in the Hindu Kush region and is the predominant activity of the seasonally nomadic Pashtuns. There are also significant numbers of cattle and swine raised; however, this production remains at a local scale and all of the products from livestock production are consumed locally.

Although the agricultural sector is the largest segment of Afghanistan's economy, it is confined to the local and regional scale because of the lack of

Economic Geography

Peering out the window of a damaged schoolhouse, one has a terrific view of a major vineyard in the Panjsher Valley. Unfortunately, due to drought and war, the vineyard has not produced any grapes for the past twenty years.

overland transportation infrastructure, such as roads, railways, and navigable rivers, which is necessary to transport goods to other markets. Additionally, the recent droughts in Afghanistan had a devastating effect on the country's 2000 food crop, hindering its ability to meet the domestic consumption requirements of its people. The lack of transportation infrastructure, coupled with the drought of 1999, forced Afghanistan to rely upon humanitarian assistance to feed its people during the 2000 and 2001 growing seasons. USAID reports, however, that Afghani food production for the 2002 season has recovered to the pre-drought level of 3.5 million tons (USAID, 2002).

INDUSTRY

In the 1970s, the industrial sector consisted primarily of natural gas exports to the former Soviet Union. Afghanistan at one point exported 70–90 percent of its output to Russia, approximately 275 million cubic feet per day (Energy Information Administration, 2001). There are two existing natural gas pipelines connecting the towns of Bagram and Shindand to Uzbekistan and Turkmenistan, respectively. Thus, a viable transportation infrastructure exists to move Afghani natural gas to the Black Sea and Western markets. Since the Soviet invasion in 1979, however, fossil fuel production has dropped to nearly zero and what little is produced in the Jowzjan Province is consumed domestically.

In an attempt to revitalize the gas and oil industry, the Taliban assumed control of the Afghan Gas Enterprise and the Afghan National Oil Company, and in 1999 started repairing the distribution pipeline near Mazar-i-Sharif (Energy Information Administration, 2001). However, United Nations Security Council Resolution 1267 froze funds and financial resources derived from property owned or controlled by the Taliban, removing Afghanistan's ability to export fossil fuel (United Nations, 1999). Consequently, the industrial sector with the greatest potential was limited by the international community due to the Taliban's refusal to abide by the Chapter VII ruling of the United Nations, which ordered the surrender of Osama bin Laden (United Nations, 1999; United Nations, 2000).

Natural gas and petroleum exports remain virtually nonexistent, but President Karzai has been consulting with the government of Uzbekistan, and the "possibility of exporting a small quantity of natural gas through the existing pipeline into Uzbekistan is reportedly being considered" (EIA 2002, 1). This initial step has great promise based on the relative location of Uzbekistan to the distribution pipeline near Mazar-i-Sharif.

At the household scale there is a small but important cottage industry in handwoven carpets that are traded across borders. This small-scale international trade provides income to the most vulnerable portion of Afghani society.

TRANSPORTATION

The transportation network in Afghanistan is inadequate to support little more than local movement. There are 21,000 km (13,041 mi) of roadway within the country, most of which is characterized as unimproved surfaces and trails, while only 2,793 km (1,846 mi) are paved and capable of supporting limited commercial wheeled vehicles (see Figure 8.2) (CIA 2001). The railroad consists of 24.6 km (15.3 mi) of track: 9.6 km (6 mi) from Gushgy, Turkmenistan, to Towraghondi and 15 km (9.3 mi) from Termiz, Uzbekistan, to Kheyrabad. There are ten airports with paved runways; however, only three are over 3,000 m (9,750 ft.) and capable of accommodating large commercial aircraft. Additionally, the United Nations banned commercial air traffic into and out of Afghanistan in 1999, and this prohibition remained in effect until 2002. Moreover, humanitarian flights were only authorized with the expressed permission of the UN Security Council (United Nations, 1999; United Nations, 2000).

PRIMARY RESOURCES

As mentioned previously, Afghanistan possesses both natural gas and crude oil. Natural gas, however, is the only fossil fuel that has been successfully exploited for export markets. Russia estimated Afghanistan's reserves to be as much as 5 trillion cubic feet of natural gas, and "proven and probable oil and condensate reserves at 95 million barrels" (Energy Information Administra-

Economic Geography

Source: John Wiegand

For the past twenty-five years, people have been displaced because of tribal conflicts, political unrest, and war. UN tents have become ubiquitous. These particular tents house refugees who were not accepted by the village in the background.

tion, 2001). In addition to the gas and oil reserves, Afghanistan is believed to hold 73 million tons of coal in the northern reaches of the country between Herat in the west and Badhakshan in the east. With these three reserves, Afghanistan not only has export potential, but it could be a significant player in the global fossil fuel market (Energy Information Administration, 2001).

Afghanistan also produces hydroelectric power. There are three dams (Kajaki, Mahipar and Breshna-Kot Dams) that are operational and produce a combined total of 94 megawatts (Energy Information Administration, 2001). This is not sufficient, however, to meet local and regional needs, prompting Herat and Andkhoy to rely upon electricity imports from Turkmenistan. These renewable and fossil fuel sources provide an economic base, which if exploited and managed successfully could propel Afghanistan into the developed world.

EVERYDAY LIFE IN AFGHANISTAN

The result of the 20 years of war and the imposition of economic sanctions has affected the lives of individual families in profound ways. First, the international conflict with the Russians during the 1980s essentially destroyed the physical infrastructure of Afghanistan and significantly retarded industrialization in the northern border region. The conflict with Russia also caused 6 million

Economic Geography

Source: John Wiegand

Water supply and water quality are two distinct factors that have a significant impact on the country's population. Outside of Bagram, a young girl retrieves water for her mother to use for cooking.

people to flee Afghanistan, effectively reducing the most affluent and skilled portion of human capital. These conditions, coupled with the internal regional power struggles spanning the 1990s have seriously reduced employment opportunities. As a result, most Afghanis resort to subsistence farming. This trend will likely continue for the foreseeable future because without skilled labor for industry and transportation improvements, families will be unable to get surplus agricultural products to market before spoilage occurs (United Nations, 2001). Thus, any surpluses are simply bartered between nomadic herders, sedentary farmers, and periodic markets. This limitation, however, may have encouraged the pursuit of an alternative economic opportunity, namely the production of opium.

ILLICIT OPIUM

The rapid growth of the opium economy in Afghanistan can be attributed to two macro and several micro factors. First, the decisions of Turkey, Iran, and Pakistan to eliminate poppy production because drugs are against the teachings of Islam, created a global deficit in heroin supply (United Nations, 2001). Secondly, the economic sanctions placed on the Taliban government prevented

Afghanistan from participating in legal international trade and denied it access to external financial institutions.

The lack of employment outside the agricultural sector within Afghanistan, and the new global demand on the black market, a market that functions independently from and is impervious to international economic sanctions, provided the economic impetus and market niche at the local scale in Afghanistan. Afghani farmers needed cash, but could not transport perishable goods to market and, even if they could, the international community denied access to licit markets. Thus, opium became a suitable cash crop because it is traded on the black market and is insulated from international sanctions. Moreover, opium has a long shelf life and can endure extended overland transportation without a loss in value. Opium also has a large value to volume ratio such that a few kilos can provide an extended family with a very comfortable annual income. To some it is not surprising that Afghanistan has evolved as the world's leading exporter of opium, providing 70–79 percent of the global supply (United Nations, 2001).

The Afghan Interim Administration (AIA) and the Islamic Transitional Government of Afghanistan (ITGA) extended the Taliban prohibition on poppy cultivation and opium production, and the current government of Hamid Karzai continues to wrestle with the issue. Poppies, however, remain a mainstay in family farming, and cultivation has actually increased from 74 tons in 2001 to 2,952 tons in 2002 (ADB 2002, 4). Compounding this is the fact is that wheat production has continued to decline and poppies are now the #1 cash crop (ADB 2002, 9). This trend has security implications for the AIA and the ITGA because funding for opposition groups and al-Qaeda continues through the illicit narcotics trade and will continue until profitable crop substitutions are developed.

SUMMARY

Despite its harsh natural environment, Afghanistan has the agricultural potential to meet the needs of its rather sparse population. Warfare, drought, earthquakes, and the lack of transportation infrastructure, however, have hindered economic development in recent years and have contributed to food shortages. Although a number of industrial minerals are present in significant quantities, the lack of investment capital, a rudimentary transportation network, and political instability have prevented their exploitation. Moreover, past political and economic sanctions have hindered relief efforts from other countries or global agencies. Based on the conditions mentioned above, Afghanistan's economic future looks bleak. However, the new government, coupled with the lifting of economic sanctions provides hope.

10

Urban Geography

Brandon K. Herl

Keypoints

- Urban centers are nodes of concentrated political and economic activity.
- Current locations of many Afghan cities relate to water availability and ancient caravan trade routes.
- Despite the resemblance in design to modern urban centers, many major Afghan cities function more as overgrown villages due to decades of war and neglect.

U rban geography, a subfield of human geography, focuses on the location, functions, and growth of urban areas. Generally, the goals of urban geographic analysis are to understand the spatial structure and organization of population centers, to examine the spatial interactions and connectivity between cities, and to explain the processes that created the observed patterns (Palka, 2001). Cities are centers of power and nodes of concentrated political and economic activity, and urban areas are just as vitally important to a rural, underdeveloped country like Afghanistan, as they are to a highly urbanized, developed country of the Western world.

As stated in previous chapters, Afghanistan has historically been a "crossroads" country. Despite its formidable challenges—the rugged aspect of the country's physical environment—the nature of its rivers and valleys and its centrality to Russia, India, China, and Iran have made it a natural meeting place and passage way for caravan traders moving to and from the Near East and Far East, as well as a conduit to funnel conquering armies. As such, the nature, structure, and location of both Afghanistan's major cities and its numerous small rural villages share a common but unique heritage. This heritage has played a key role in the site and situational factors that determine any city development (Getis et al., 2001).

This chapter focuses initially on the commonalities between Afghanistan's population centers and the physical aspects of their sites. We subsequently examine the situational attributes that influence each city's general layout and growth. We conclude this chapter with a general description of the country's local villages (that accommodate nearly two-thirds of the population) and a brief focus on its major cities, five of which (Kabul, Kandahar, Herat, Mazar-i-Sharif, and Jalalabad) will be discussed in detail.

COMMON POPULATION CENTER CHARACTERISTICS

Spawned by early land-based trade route economics, tempered by the need to support life in a harsh, unforgiving land, and forged by the necessity to protect themselves from numerous invading armies and constantly marauding parties of bandits, Afghan population centers often appear to be an unusual cross between a desert fortress stronghold and a preindustrial age merchant bazaar. While each city and village is generally self-sufficient, most are roughly interconnected in a linear fashion that reveals some of the earliest land trade routes (Edwards, 2001). This connectivity is *not* an accidental occurrence.

The initial selection of a village or town is usually related to aspects of its *site*. As a geographic concept, site refers to the internal aspects of a place, especially the physical characteristics. In Afghanistan, most of the cities and villages have always relied heavily upon agriculture and trade since their founding. These areas became central places where farmers and herders could gather together for protection, as well as congregate to buy or sell goods or acquire special services industries. In this respect, Afghanistan's villages, towns, and cities are not any different from urban places of the developed world. The former provide goods and services to its residents and people of the surrounding hinterland. Much of rural Afghanistan today still is reliant upon this ancient arrangement and all its major cities still have a heavy agricultural link (Edwards, 2001, *Afghanistan Country Review*, 2001).

In this arid part of the world, water is key to survival. No village, farmer's field, or herder's flocks could exist without a reliable, easily accessible water source. Consequently, all major Afghan cities and villages are located on relatively flat ground on or near perennial rivers. As a general rule, the size of the population center is limited by the baseflow of the river unless machinery is available to drill for groundwater. Locales sporting high volumes of open or flowing water and positioned on flat, fertile ground support larger populations. In other places where the river's flow is not as great and the land is irregular, or not as well suited to agriculture, town growth is extremely limited (Edwards, 2001). The general locations where abundant water and gentle relief exist are often where channelized rivers converge and exit rugged mountain areas into valleys that merge into flatlands. It is no coincidence that Afghanistan's largest cities are positioned at these key locations (see Figure 10.1).

Smaller cities and villages tend to be spaced roughly 100 km (62.5 mi) apart from each other in the flatter, less rugged areas of the country. These locations are points where historically, caravans would have had to acquire additional water during their long treks between the Far East and Near East.

Urban Geography

Figure 10.1 Afghanistan's Road Network

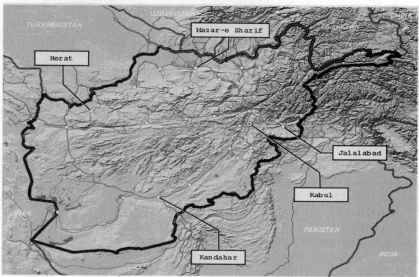

Source: Brandon Herl

A typical caravan would travel at walking speeds of about 3–5 km per hour (2–3 mi per hour) for up to ten hours a day. This rate of movement enabled a caravan to travel roughly 30–50 km (19–31 mi) per day in open country. A caravan's money-making cargo included spices, silks, and other exotic items, but not water. Precious space and weight were devoted to high-payoff cargo and were rarely sacrificed for hauling water. As a result, caravans would often carry only a two- or three-day supply of water for the entire group of pack animals, merchants, and armed escort guards (Marco Polo as edited by Marsden & Wright, 1948).

At caravan travel rates of 30 km a day, a caravan sought a water supply point every three days, or at distances of 90 km (56 mi). When traveling at the faster rates of 50 km per day, heavily burdened animals consume more water and deplete supplies quicker, forcing a caravan leader to either make more frequent water stops or carry more water. Based upon this economic decision, taking on more water every two days yields a planned stop every 100 km (63 mi)—the same distance between most Afghan cities. In rougher terrain, such as in the mountain region of Afghanistan's interior, these optimum daily travel distances were much less, so caravans interacted with more (but smaller) villages.

Although villages and towns are initially located based on aspects of their site, they tend to grow over time based on aspects of their *situation*. The latter concept refers to the external aspects of a place, and emphasizes the connection of the place relative to other places or events. Stemming from the caravan scenario is a second characteristic that generally defines the layout and structure of many Afghan cities and villages. Caravans needed more than a place to rest and replenish water supplies for animals and people; they also needed a place to conduct business and protect their wares. These requirements helped to shape the layout of many of Afghanistan's past and present-day villages and towns, as well as provided the central core structure for the country's larger cities.

Approaching a village in the manner of a traveling merchant is a good way to visualize how the layout of a typical Afghan town may have evolved. Imagine that after two or three hard days of sweating and wiping the ever-present dust out of his stinging eyes, the merchant sees that his hard-packed road leads into a fairly flat, open area. He begins to smell the lush, green field crops near the river's edge. He sees numerous watering areas on the outskirts of town and looks to find a large open area near water, but not next to, the cool green fields. He camps in these open areas and takes a small part of his caravan into the main city.

As he follows a road leading into a large, open, central area easily accessible from the main route of trade travel, he observes that this road passes through irrigated fields and trees, until he is confronted with large, flat-roofed, mud-brick buildings with exterior walls devoid of windows. Some buildings are larger than others, but most are only one to two stories high. These buildings serve the dual purpose of protecting the inner city, as well as providing needed shelter from midday sun or wind-driven sandstorms.

As the merchant continues into the small but rectangular city, he is funneled toward the central trading bazaar, passing by dwellings occupied by workers until he reaches the main city square, which doubles as a large, open-air trading market. He notes many temporary stores fashioned in tents or on carts set up by other traveling merchants or locals who live in the city's outskirts. The best shops are along the boundaries of the central market square, where the permanent shops also double as living quarters for the more established traders and service providers.

This area also is near a larger, more permanent structure resembling the outer city buildings, but made of thicker walls and often of stone. Here is the central governmental center that houses the town's leadership and police. He also recognizes another more permanent building near this central area, the mosque.

The above scenario may appear to be a scene straight out of Marco Polo's fifteenth-century travels, but many of Afghanistan's smaller villages still resemble this layout, especially in the country's rugged interior and more isolated northern areas. While the major Afghan cities are at times quite large and sprawling as compared to their more humble ancestral roots, they still tend to follow this romanticized structural pattern (Marco Polo, 1948; Edwards, 2001).

MODERN URBAN CENTERS

Despite the semblance of traditional layouts, the complexions of Afghanistan's largest cities have been altered by modern developments in transportation,

Urban Geography

Figure 10.2 Russian Topograghic Line Map Extract of Kabul

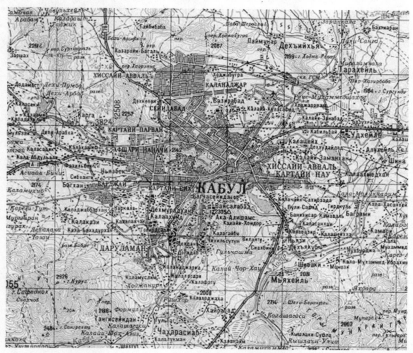

Source: Soviet General Staff Map I-42-XVI, 1985

communication, construction practices, and economic activities. In an effort to modernize, many old city districts have been razed to make way for more modern buildings, wider truck-capable roads, and other general infrastructure improvements, although some cities have preserved many of the older monuments celebrating Afghanistan's history (*Britannica Online*, 2001). The ages, locations, layouts, and experiences of each major city contribute to unique and distinguishing characteristics that are clearly visible in the cultural landscape (Palka, 2001). A general examination of five major Afghan cities (Kabul, Kandahar, Herat, Mazar-i-Sharif, and Jalalabad) scattered throughout the country demonstrates this point.

It is noteworthy that despite many governmental modernization programs (starting in the 1970s and continuing throughout the Soviet occupation in the 1980s), most cities have fallen victim to the ravages of war. Moreover, infighting has continued during the occupation period and throughout the country's civil war over the past two decades. Consequently, virtually all of the major cities have experienced significant damage to their infrastructure

Figure 10.3 Russian Topographic Line Map Extract of Kandahar

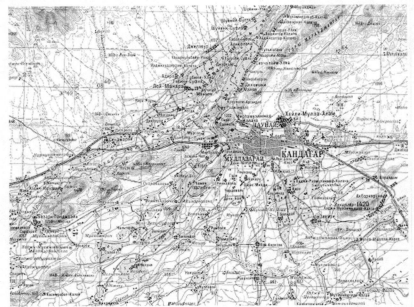

Source: Soviet General Staff Map H-41-VI, 1985

and many of the urban landscapes exhibit various types of visual blight and decay.

Kabul is Afghanistan's capital and its largest city with a population of around 1.5 million people (*Afghanistan Country Review*, 2001; *Britannica Online*, 2001; *CIA World Factbook*, 2001). The city is located within Kabul Province along the Kabul River in the eastern part of the country, approximately 225 km (140 mi) from the border with Pakistan. It is recognized as the country's *primate city* and serves as the economic, governmental, and cultural center. Figure 10.2 is extracted from a Russian topographic military map and serves to illustrate its location in a flat, triangular-shaped valley near a river and along a major road. Also note the interconnected irrigation canals located in the lesser-developed eastern outskirts of the city.

To gain an appreciation of Kabul's relative importance to the nation, look closely at the road infrastructure leading to and from the city. The road to the north leads into the Hindu Kush Mountains and toward Uzbekistan and Tajikistan, both former Soviet republics. The southern road leads toward the crossroad city of Kandahar, which provides passage further south through Pakistan toward the Indian Ocean or westward toward Iran. The road heading east goes toward Jalalabad and the Khyber Pass, which accesses Peshawar and Islamabad in Pakistan. In addition to being the country's central economic, govern-

Figure 10.4 Russian Topographic Line Map Extract of Herat

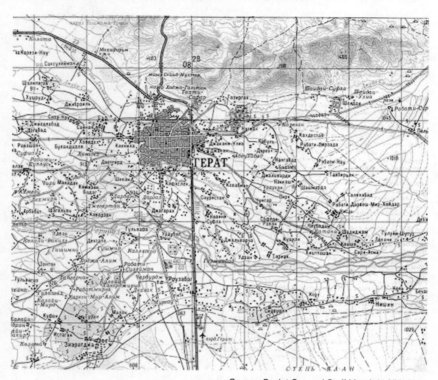

Source: Soviet General Staff Map I-41-XV, 1985

mental, and cultural center of gravity, it is also a critical focal point that maintains links and ties with nearly all areas of the country.

Kabul has been in existence for over 3,500 years, mainly due to its location on the ancient travel routes, and consequently it has changed hands many times. As such, the complex cityscape reveals cultural imprints fashioned by conquering armies and rulers, although current inhabitants, mostly speakers of Dari (Persian) with a large minority of Pashtuns, continue to modify the cultural landscape to suit their own needs (*Britannica Online*, 2001).

Kandahar (Qandahar) is Afghanistan's second-largest city with a population of over 225,000 people, mainly of Pashtun descent. Located within Kandahar Province in the southeastern part of the country, the city is positioned approximately 100 km (63 mi) from the Pakistan border. While not a major industrial manufacturing center like Kabul, Kandahar is a major commercial and trade center due mainly to its location between Herat and Kabul, as well as its situation along the roads that lead westward into Iran and southward into Pakistan. This strategic trade location has been a bane to Kandahar. The city has

Figure 10.5 Russian Topographic Line Map Extract of Mazar-i-Sharif

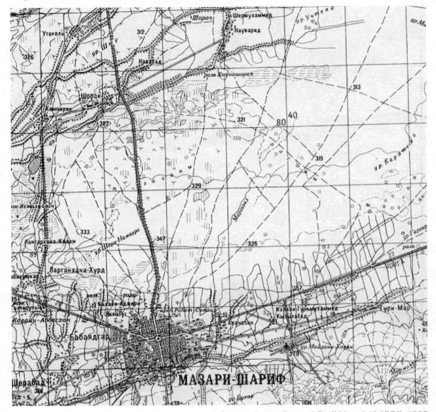

Source: Soviet General Staff Map J-42-XXVI, 1985

changed hands nearly every time a new conquering army has moved through Afghanistan. As such, it exhibits ruins and influences of nearly every conquering army to travel through this part Central Asia, including Alexander the Great (*Afghanistan Country Review*, 2001; *Britannica Online*, 2001; *CIA World Factbook*, 2001).

While a modern city in many respects, Kandahar still clings to its old traditional founding and structure, and in many respects seems much more like an oversized Afghan village rather than the country's second-largest city. Despite the creation of a more modern city on the outskirts of the traditional "old-city" area, the land surrounding Kandahar is irrigated and farmed and is dotted with orchards and vineyards, as illustrated in Figure 10.3. Industry in the city still revolves around agriculture (especially food processing), as well as natural textile manufacturing (most notably wool).

Urban Geography

Figure 10.6 Russian Topographic Line Map Extract of Jalalabad

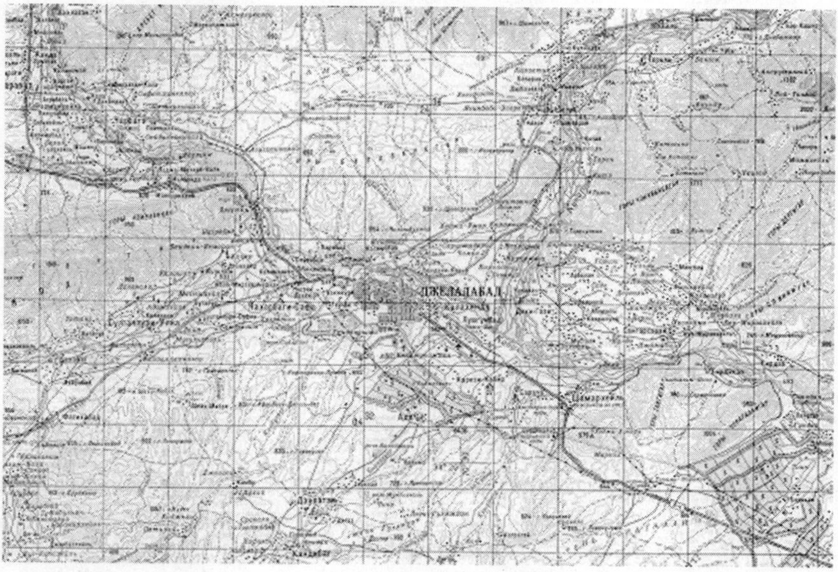

Source: Soviet General Staff Map I-42-XVII, 1985

Herat, the country's third-largest city with over 180,000 people, is western Afghanistan's economic center (see Figure 10.4). The city is located on the Hariud River within Herat Province, approximately 110 km (69 mi) from the border with Iran. The surrounding region has extremely fertile lands and is home to the country's most densely populated agricultural regions. Like many other Afghan cities, Herat, too, has a history of multiple conquests and is surrounded by the ruins of many other old cities. Herat also incorporates a moderate fur trade, although it would be stretching the point to consider it one of the city's major industries. Of special note is Herat's population, which is mainly Tajik, Turkmen, and Uzbek, much different than the Dari- (Persian-) speaking groups in Kabul (*Afghanistan Country Review*, 2001; *Britannica Online*, 2001; *CIA World Factbook*, 2001).

Mazar-i-Sharif (see Figure 10.5) is another traditionally agricultural city that primarily produces cotton, grains, and fruits. Located within Balkh Province in the north-central part of the country, Mazar-i-Sharif's population is over 130,000, consisting mainly of Uzbeks, Tajiks, and Turkmens due to its proximity to Uzbekistan, which is 35 km (22 mi) further north. The town's name literally means "tomb of the saint" since the caliph Ali, the son-in-law of the Prophet Muhammad, reportedly lies in the blue-tiled mosque that marks his tomb. This northern city is highly regarded as a special holy place by all believers of Islam, with the distinct exception of Shi'ite Muslims.

Jalalabad (Jalakot) (see Figure 10.6) is a smaller city that is located roughly 170 km (106 mi) east of Kabul. The city is located along the Kabul River within Nangarhar Province and is positioned 50 km (31 mi) west of the Pakistan border. Jalalabad supports a population of nearly 60,000 people, mainly of Pashtun descent. The city conforms to the typical Afghan city layout, surrounded by large irrigated plains, and boasts an industrial base that is mainly related to agriculture or food processing. Jalalabad is roughly midway between Kabul and Peshawar in Pakistan. Most importantly from a military perspective, Jalalabad is the largest Afghan city near the critical Khyber Pass and also dominates the entrances to the Laghman and Kunar valleys (*Afghanistan Country Review*, 2001; *Britannica Online*, 2001; *CIA World Factbook*, 2001).

SUMMARY

Afghanistan's major cities have been heavily influenced by their physical location, as well as by their long, often complicated histories. Repetitive conquests have contributed to a mixed urban landscape for many major cities, with aspects of the cultural imprint reflecting the city's former role as a key point along an ancient caravan trade route. While there are vast differences between the size and lifestyles of small rural villages and major cities such as Kabul and Mazar-i-Sharif, it is interesting to note that even modern influences have failed to radically change city structure and function.

11

Population Geography

Dennis D. Cowher

Key Points

- Afghanistan has an ethnically diverse population of 26 million people.
- The country's high rate of natural population growth and the return of refugees from neighboring countries may continue to hinder efforts to build a stable government.
- Short life expectancy and high infant mortality rates among the population are indicative of an underdeveloped country.

Demography, or the study of characteristics of human populations, is an interdisciplinary undertaking. Geographers approach the study of population with a unique perspective. They study population to understand the spatial distribution of Earth's people. Geographers are also interested in the reasons for, and consequences of, the distribution of population from the local to global scales. While historians study the evolution of demographic patterns and sociologists address the social dynamics of human populations, geographers focus special attention on the spatial patterns of human populations, the implications of such patterns, and the reasons for them. Using many of the same tools and methods of analysis as other population analysts, geographers think of population within the context of the places that populations inhabit.

Demography, or the systematic analysis of the numbers and distribution of people, enables the analyst to explore the interrelationships and interdependencies between humans and places. Given geography's emphasis on different people and places, the discipline offers unique opportunities to examine the population distribution and characteristics of the state of Afghanistan. This chapter examines the population geography of Afghanistan and answers three important questions. First, what are the spatial patterns of human population in Afghanistan? Second, what are the causes and consequences of such a population distribution? And third, what are the composition and growth characteristics of the Afghan population?

POPULATION DISTRIBUTION

Afghanistan has many ethnic groups and these groups have a distinct spatial organization within the country. The Pashtuns, the largest group and an estimated 38 percent of the total population, live predominantly in the east and south. Tajiks, who account for a quarter of the population, are the second-largest ethnic group and live mostly in the northern region of the country. Other groups include the Uzbeks and Turkmen, who live in the north central region, and the Hazara, who live in the central region. There are smaller numbers of Baluchi, Brahui, Nuristani, Aimak, Qizilbash, and Kyrgyz (CIA *World Factbook*, 2001).

According to the U.S. Census Bureau, there were approximately 26 million people living in the country in 2000. It is important to note that Afghanistan has one of the lowest rates of urbanization in the world. About 80 percent of the population is considered rural (CIA *World Factbook*, 2001). This is not surprising, given that approximately 70 percent of the people earn their living from agriculture. There are many environmental and physical factors that have important influences on population distributions and concentrations. Some of these factors are degree of accessibility, topography, soil fertility, climate and weather, water availability and quality, and type and availability of other natural resources. Other factors are also crucial; foremost are the country's political, cultural, and economic experiences and characteristics. For example, Afghanistan's capital city, Kabul, is the country's largest, with a population of about 1.5 million. Afghanistan's second largest city is Kandahar. Located in southern Afghanistan on a fertile, irrigated plain, Kandahar is the chief commercial center of the country. The leading products of the province are fruit, grain, tobacco, silk, cotton, and wool. The city itself has fruit processing and canning plants and textile mills. The nation's third largest city, Mazar-i-Sharif, is a commercial center for the northern region of the country and is an important pilgrimage destination. The fifteenth-century mosque in the city is said to contain the tomb of the caliph Ali, son-in-law of Muhammad.

There is an uneven spatial distribution of people in Afghanistan. Some provinces have agglomerations of people while others are sparsely populated. Table 11.1 provides data on area and population for 30 of 32 provinces.

Afghanistan's rural population is approximately 20 million. With the current political turmoil and food shortage in the countryside, city populations could grow rapidly, causing even more health and nutritional problems. Cities have important administrative and military command functions and provide a refuge for impoverished rural groups. Additionally, many Afghans are attempting to leave the country, emigrating to Iran or Pakistan. Afghanistan's rural population also includes an estimated 2 million nomads, most of whom are Pashtuns, who move from winter grazing sites in the valleys and plains to the west, northwest, and southwest of the Hindu Kush to summer pastures in the Hazarajat and in Badakhshan to the northeast (Lieberman, 1980).

Population Geography

Table 11.1 Afghanistan's Population by Province

Administrative division	Capital	Area (sq km)	Population 2000
Badakhshan	Feyzabad	44,059.3	923,144
Badghis	Qal'eh-ye Now	20,590.6	413,254
Baghlan	Baghlan	21,118.4	944,407
Balkh	Mazar-i-Sharif	17,248.5	1,113,620
Bamian	Bamian	14,175.3	475,750
Farah	Farah	48,470.9	483,891
Faryab	Meymaneh	20,292.8	1,070,072
Ghazni	Ghazni	22,914.6	1,245,589
Ghowr	Chaghcharan	36,478.8	810,213
Helmand	Lashkar Gah	58,583.7	1,034,672
Herat	Herat	54,778.0	1,519,882
Jowzjan	Sheberghan	11,798.3	1,193,643
Kabol	Kabol	4,461.6	2,838,587
Kandahar	Kandahar	54,022.0	1,159,095
Kapisa	Mahmud-e Raqi	1,842.1	529,855
Khost		4,151.5	--
Konar	Asadabad	4,941.5	493,962
Kondoz	Kondoz	8,039.7	1,253,565
Laghman	Mehtar Lam	3,842.6	627,050
Lowgar	Pule Alam	3,879.8	424,621
Nangarhar	Jalalabad	7,727.4	1,451,917
Nimruz	Zaranj	41,005.4	199,576
Nurestan		9,225.0	--
Oruzgan	Tarin Kowt	30,784.5	736,805
Paktia	Gardiz	6,431.7	930,768
Paktika	Sharan	19,482.4	493,691
Parvan	Charikar	9,584.4	918,898
Samangan	Aybak	11,261.9	594,751
Sar-e Pol	Sar-e Pol	15,999.2	--
Takhar	Taloqan	12,333.0	845,393
Vardak	Meydan Shahr	8,938.1	729,982
Zabol	Qalat	17,343.5	349,960
Total		645,806.5	25,806,613

Source: Data from Oak Ridge National Laboratory

For the non-nomadic, largely agricultural segment of the rural population, density of settlement is greatest in eastern Afghanistan in the intensively cultivated plains and valleys formed by the Kabul River and its tributaries (Lieberman, 1980). A second region of significant rural settlement includes the lower valleys of the Kunduz and Khanabad Rivers (provinces of Takhar and

Kunduz), which drain into the Amu Darya (Lieberman, 1980). There are few inhabitants living in the provinces located in the deserts and mountainous wastelands of western and central Afghanistan.

POPULATION DENSITY

Another way to examine the population is in terms of density, a numerical measure of the relationship between the number of people and some other unit of interest expressed as a ratio. For example, crude density (sometimes referred to as arithmetic density) is probably the most common measurement of population density. Crude density is the total number of people divided by the total land area.

Afghanistan has a population of 25,889,000 within a total land area of 251,825 sq mi (652,225 sq km). Therefore, the country's population density is equal to 106 persons per sq mi (or 41 persons per sq km) (U.S. Census Bureau, 2001). For the sake of comparison, Afghanistan is slightly smaller than the state of Texas. Texas has a population of 20,044,141 living on 267,277 sq mi (692,244 sq km). Thus, Texas has a population density of 77 persons per sq mi (or 30 persons per sq km). We can conclude from this data that Afghanistan is 37 percent more densely populated than the state of Texas. Another important concern is that this density does not take into account the limited amount of arable land in Afghanistan. According to the CIA, only 12 percent of the land in Afghanistan is considered arable (CIA *World Factbook*, 2001). An additional 46 percent is considered permanent pastures (CIA *World Factbook*, 2001). This means there is even greater competition among the varied ethnic groups for the land that is capable of producing food.

POPULATION COMPOSITION

In addition to exploring patterns of distribution and density, population geographers also examine population in terms of its composition, that is, in terms of the subgroups that constitute it. Understanding population composition enables analysts to gather important information about population dynamics. For example, knowing the composition of a population in terms of the total number of males and females, number of proportions of old people and children, and number and proportion of people active in the workforce, provides valuable insights into the ways in which the population behaves.

The most common way for geographers to represent graphically the composition of the population is to construct an *age-sex pyramid* (or more properly referred to as a population profile), which is a representation of the population based on its composition according to age and sex cohorts. Usually, males are portrayed on the left side of the vertical axis and females to the right. Age categories are ordered sequentially from the youngest, at the bottom of the pyramid, to the oldest, at the top. By moving up or down the pyramid, one can

Population Geography

Figure 11.1 Age-Sex Pyramid for Afghanistan, 2000

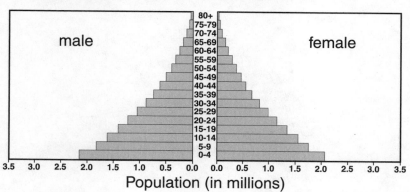

Population (in millions)

Source: Data from Oak Ridge National Laboratory

compare the opposing horizontal bars to assess differences in frequencies for each age group. A cohort is a group of individuals who share a common temporal demographic experience. A cohort is not necessarily based on age, however, and may be defined according to criteria such as time of marriage or time of graduation.

Age-sex pyramids can reveal the important demographic implications of war or other significant events. Moreover, age-sex pyramids can provide information necessary to assess the potential socioeconomic impacts that growing or declining populations might have on the country. Figure 11.1 is an age-sex pyramid for Afghanistan for the year 2000.

Afghanistan's population profile reveals a high proportion of dependent children, ages 0 to 14, relative to the rest of the population. The considerable narrowing of the profile toward the top indicates that the population has been growing very rapidly in recent years. The pyramidal shape of Afghanistan's profile is typical of countries with high birth rates and relatively low death rates. Serious implications, however, are associated with this type of profile. First, in the absence of high productivity and wealth, resources are increasingly stretched to their limit to accommodate even elemental schooling, nutrition, and health care for the growing number of dependent children. Furthermore, when these children reach working age, a large number of jobs will need to be created to enable them to support themselves and their families. Additionally, as they form their own families, the sheer number of women of childbearing age will almost guarantee that the population explosion will continue. This scenario will become a reality unless the country's government takes strong measures. The range of initiatives include intensive birth control campaigns, improved education and employment opportunities outside the home for women, and awareness campaigns to modify cultural norms that place a high value on large family size. Given the cultural and political information that has

Population Geography

Figure 11.2 Projected Age-Sex Pyramid for Afghanistan, 2025

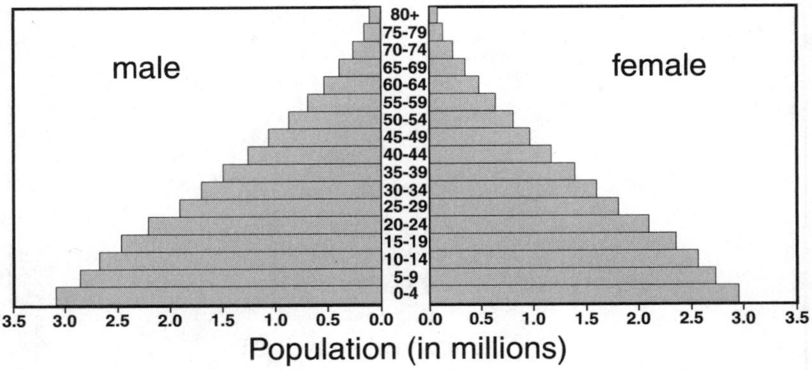

Population (in millions)

Source: U.S. Census Bureau, International Database

already been discussed in previous chapters, it is very unlikely that the current government will be able to implement the necessary measures to reduce the natural growth rate of its population in the foreseeable future. Therefore, we can expect the population of Afghanistan to continue growing rapidly, unless war or famines occur in the future. Based on current trends, the Census Bureau can project the country's age-sex pyramid for 2025 (see Figure 11.2).

A critical aspect of the population pyramid is the *dependency ratio*, which is a measure of the economic impact of the young and old on the more economically productive members of the population. In order to assess this relation of dependency in a particular population, geographers will typically divide the total population into three age cohorts. The youth cohort consists of those members of the population who are less than 15 years of age and generally considered to be too young to be fully active in the labor force. The middle cohort consists of those members of the population aged 15 to 64 who are considered economically active and productive. Finally, the old-age cohort consists of those members of the population aged 65 and older who are considered beyond their economically active and productive years. By dividing the population into these three groups, it is possible to obtain a measure of the dependence of the young and old upon the economically active, and the impact of the dependent population upon the independent.

Afghanistan has a very large dependency ratio. Forty-four percent of the population is under age 15 or over age 65. From Figure 11.1, it is clear that the main reason for this high dependency ratio is the large number of children in Afghanistan's population. There are nearly 11 million children in Afghanistan under the age of 15. This is 42 percent of the population. In contrast, children under age 15 in the United States only account for 21 percent of the total population.

BIRTH RATES AND DEATH RATES

The demographic statistics of Afghanistan presented in this section are consistent with the patterns in other predominantly rural and economically developing Islamic countries. The *crude birth rate* (CBR) is the total number of live births in a year for every thousand people in the population. The crude birth rate for Afghanistan in the year 2000 was 42 (U.S. Census Bureau, 2001). To put this number into perspective, consider that it is three times the corresponding figure in the United States, which is 14. Although the level of economic development is a very important factor shaping the CBR, other, often equally important influences also affect CBR. Afghanistan, in particular, is influenced by women's educational achievement, religion, social customs, diet and health, as well as war and political unrest.

The crude birth rate is only one of the indicators of fertility. Another indicator used by population experts is *total fertility rate* (TFR), which is a measure of the average number of children a woman will have throughout her childbearing years, generally considered to be ages 15 through 49. Whereas the CBR indicates the number of births in a given year, the TFR is a more predictive measure that attempts to portray what birth rates will be among a particular cohort of women over time. A population with a TFR of slightly higher than 2.0 has achieved replacement level fertility. This means that birth rates and death rates are approximately balanced and there is stability in the population. The TFR for Afghanistan is 5.9, whereas in the United States and many other developed countries it is 2.1. (U.S. Census Bureau).

Closely related to the TFR is the doubling time of the population. The *doubling time*, as the name suggests, is a measure of how long it will take the population of an area to grow to twice its current size. To compute a country's doubling time, we simply divide the number 70 by the rate of natural increase. In the case of Afghanistan, the rate of natural increase is 2.4. If we divide 70 by 2.4, we get a period of 29 years. The idea that Afghanistan's population will double to 52 million in the next three decades is troubling, given the region's political instability and inability to feed its current population.

Countering birth rates and also shaping overall population numbers and composition is the *crude death rate* (CDR), the ratio between the total number of deaths in one year for every thousand people in the population. Crude death rates often reflect levels of economic development. Mortality is high in Afghanistan for infants, children, and adults alike. The CDR is 18, which is twice as high as the U.S. CDR of 9.

Death rates can be measured for both sex and age cohorts and one of the most common measures is the *infant mortality rate*. This figure is the annual number of deaths of infants less than one year of age compared to the total number of live births for that same year. The figure is usually expressed as number of deaths during the first year of life per 1,000 live births. The infant mortality rate has been used by researchers as an important indicator both of a

Population Geography

Source: Eugene Palka

Young boys in the village of Charikar appear happy and healthy. Disease, a lack of health care, and warfare, however, have prevented many children from living a full life.

country's health care system and the general population's access to health care. Afghanistan's infant mortality rate is a staggering 147 deaths per 1,000 live births (CIA *World Factbook*). To put this figure into perspective, compare it to Europe's average infant mortality rate of just 7.

Related to infant mortality and the crude death rate is *life expectancy*, the average number of years an infant newborn can expect to live. Infants born in Afghanistan in the year 2000 can expect to live an average of 46 years, while U.S. infants born in the same year can expect to live 77 years.

High infant mortality rates and low life expectancy levels exist even in the capital city of Kabul, which has the highest educational levels and best medical facilities in the country. These statistics testify to the inadequacy of the public health effort. Analyses of the causes of death in Afghanistan emphasize the effects of malnutrition (as a direct and contributing factor) and infectious diseases such as diphtheria, tetanus, and pneumonia on the health of children; adult mortality has been attributed to many of the respiratory and gastrointestinal diseases that affect children. In addition, tuberculosis and malaria, a debilitating disease, continue to have a high incidence among adults in Afghanistan (Lieberman, 1980).

MOBILITY AND MIGRATION

In addition to the population dynamics of death and reproduction, the movement of people from place to place is a critical aspect of examining population geography. Mobility is the ability to move from one place to another, either permanently or temporarily. We have already discussed the nomadic characteristics of the Afghan population. Millions of rural Afghans, mostly Pashtuns, move with the seasons for better grazing opportunities for their herds of sheep and goats. The second way to describe population movement is in terms of *migration*, which is a move to a new location with the intention of being permanent. Migrants intend to permanently change their place of residence—where they sleep, store their possessions, and receive legal documents for the foreseeable future.

Migration has two forms, emigration and immigration. Emigration is migration out from a location; immigration is migration into a location. A decision to migrate stems from a perception that somewhere else is a more desirable place to live. People may hold very negative perceptions of their current place of residence or very positive perceptions about the attractiveness of somewhere else. Negative perceptions about their place of residence that induce people to move away are *push factors*, whereas *pull factors* attract people to a particular new location.

Migration from Afghanistan is basically resulting from three push factors: political, economic, and environmental. Refugees are people forced to migrate from a particular country for political reasons. The United Nations defines political refugees as people who have fled their home country and cannot return for fear of persecution because of their race, religion, nationality, and membership in a social group, or political opinion (Rubenstein, 1996).

As a result of the former Soviet Union's invasion of Afghanistan in 1979, more than 5 million Afghans fled to refugee camps set up in neighboring Iran and Pakistan. Because of a very high natural increase rate—an average of 2.6 percent since 1979—the population in the refugee camps has swelled to more than 6 million (Rubenstein, 1996). By the mid-1990s, more than 3 million Afghans lived in tents or mud huts set up in 250 camps in Pakistan. The largest number live near the town of Peshawar, in northern Pakistan. Peshawar is situated near the eastern end of the Khyber Pass, the major land route through the mountains between Afghanistan and Pakistan. Other Afghan refugees settled in camps in Pakistan's Baluchistan and Punjab provinces. More than 2 million Afghan refugees migrated westward to Iran, primarily to the border cities of Mashhad, Birjand, and Zahedan, as well as the capital of Tehran.

Since the withdrawal of Soviet troops in 1989 and the collapse of the Soviet-installed government in 1992, several thousand refugees have returned home, trading the security of the camps for the possibility of reclaiming their farms. The UN had issued ration books to the refugees so that they could obtain food while living in the camps and provided each returning family with about 300 kilograms of wheat and the equivalent of $150 to pay for transportation (Rubenstein, 1996).

Population Geography

Source: John Wiegand

Young children from the village of Charikar give a thumbs-up gesture to American troops passing by the village.

     Throughout most of the 1990s, rival ethnic groups fought for control of the government. This prevented many of the refugees from returning home. When the predominantly Pashtun Taliban gained control of Kabul and most of the country, additional ethnic minorities emigrated. In addition to the political push factors, economic and environmental push factors continue to provide reasons for Afghans to emigrate to the camps in Iran and Pakistan. The lack of job opportunities and severe drought in recent years has forced many rural Afghans to flee their homeland to pursue opportunities outside the country. Since the overthrow of the Taliban, the migration trend has reversed, although most estimates are difficult to confirm. The United Nations refugee agency has begun repatriating thousands of Afghan refugees from approximately 200 camps in Pakistan. Many of the refugees have been in Pakistan for up to two decades. The UN plans to get 600,000 refugees back into Afghanistan this year, a significant number of Pakistan's Afghan refugee total of 2 million (Anderson, 2003).

SUMMARY

The discipline of geography brings a unique spatial perspective to the scientific study of population. This chapter has employed this perspective to examine the population distribution and characteristics of Afghanistan. Humans are not

Source: Eugene Palka

A local truck driver and helping hands coordinated with a military police soldier during their delivery to Bagram airbase.

distributed uniformly across Afghanistan. The physical geography of the country and the region's harsh climate influence the current distribution of human habitation. Although 80 percent of the population is rural, there are major population concentrations in the country's primary cities. The main concentration is around the capital of Kabul. Afghanistan's demographic statistics paint a bleak picture of current and future prospects for the country. If natural growth rates are not lowered by aggressive policies, Afghanistan will continue to suffer from famines and human misery on a massive scale.

The segment of population that will suffer the most is the children of Afghanistan, who make up over 40 percent of the current population. The crisis between Taliban government and the U.S.-led coalition provided another push factor that changed the population geography of the country. Many Afghans left the country in advance of U.S. intervention during the fall of 2001. Others were forced to serve in the Taliban's military force. Many have already returned to the country with their hopes resting on the shoulders of a new government, humanitarian relief, and relative stability.

12

Medical Geography

Patrick E. Mangin

Key Points

- Afghanistan harbors numerous endemic and epidemic diseases.
- The country lacks the physical and social infrastructure to confront disease prevention and treatment, with the most serious deficiencies being the lack of water treatment systems and limited access to health care.
- Rugged terrain, periodic droughts, and warfare have limited agricultural output, contributing to malnutrition and undernutrition.

Medical geography is "the application of geographical perspectives and methods to the study of health, disease and health care" (Johnson, 1996). Medical geography incorporates two broad areas of study. The first concerns the spatial *ecology of disease* and geographical aspects of the health of populations. The second emphasizes the geographical organization of health care. Medical geography retains associations with other disciplines outside of geography, reflecting the complexity of most health-related problems and the need to examine them from a multidisciplinary perspective.

An important consideration in assessing a region's environmental health hazards is the health status of the indigenous population. This chapter will analyze the overall health of the Afghan people, focusing on the distribution of disease, nutrition, and health care.

THE SOVIET EXPERIENCE IN THE 1980s

The Soviet experience in Afghanistan provides lessons that can be applied to military operations or work by nongovernmental organizations (NGOs) conducting humanitarian assistance. The following excerpt from a paper published by the Foreign Military Studies Office provides a dramatic account of the force protection challenge in Afghanistan.

> Of the 620,000 Soviets who served in Afghanistan, 14,453 were killed or died from wounds, [whereas] accidents or disease [were] a modest 2.33

percent of the total who served. However, the rate of hospitalization during Afghanistan service was remarkable. The 469,685 personnel hospitalized represented almost 76 percent of those who served. Of these, 53,753 (11.44 percent) were wounded or injured. Fully 415,932 (88.56 percent) were hospitalized for serious diseases. In other words, 67 percent of those who served in Afghanistan required hospitalization for a serious illness. These illnesses included 115,308 cases of infectious [Type A] hepatitis and 31,080 cases of typhoid fever. The remaining 269,544 cases were split between plague, malaria, cholera, diphtheria, meningitis, heart disease, shigellosis (infectious dysentery), amoebic dysentery, rheumatism, heat stroke, pneumonia, typhus and paratyphus. (Grau, 97)

THE TRIANGLE OF HUMAN ECOLOGY

A useful framework for analyzing the impact of health-related issues in a place is provided by the *triangle of human ecology* (Figure 12.1). Three vertices form the triangle: population, behavior, and habitat. These vertices enclose the state of health (Meade et al., 1988).

Figure 12.1 The Triangle of Human Ecology

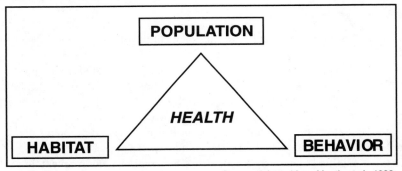

Source: Adapted from Meade et al., 1988

Habitat is that part of the environment within which people live. It includes houses, workplaces, settlement patterns, recreation areas, and transportation systems. *Population* considers humans as the potential hosts of various diseases. Factors affecting and yet characterizing the population include nutritional status, genetic resistance, immunological status, age structure, and psychological and social concerns. *Behavior* includes the observable aspects of the population and springs from cultural norms. It also impacts on those who come into contact with disease hazards and whether or not the population elects other alternatives. *Health* is a state of complete physical, mental, and social well-being, and not merely the absence of disease. Health is a continuing property that can be measured by an individual's ability to rally from a wide range and considerable amplitude of *insults*. (Palka, 2001)

Physical insults could refer to air quality, temperature, humidity, light, sound, atmospheric pressure, and trauma. Physical insults unique to Afghanistan include the stress of extreme annual and diurnal temperatures and high altitudes. *Chemical insults* include pollen, asbestos, various pollutants, smoke, or even food (Palka, 2001). Afghanistan's urban air pollution and use of pesticides and fertilizers that are banned in the United States are examples of chemical insults (DIA, AFMIC, 2001). Infectious insults include virus, bacteria, fungi, and protozoa. Infectious insults cause debilitating *endemic* and *epidemic* *diseases* in Afghanistan.

WATER-BORNE DISEASES

Sanitation is extremely poor throughout the country, including major urban areas. Local food and water sources (including ice) are heavily contaminated with pathogenic bacteria, parasites, and viruses to which most U.S. residents have little or no natural immunity.

"Diarrheal diseases (cholera and dysentery) can be expected to temporarily incapacitate a very high percentage of personnel within days if local food, water, or ice is consumed. Hepatitis A, typhoid fever, and hepatitis E can cause prolonged illness in a smaller percentage (DIA, AFMIC, 2001).

Hepatitis is not normally endemic to Afghanistan; however, it took a severe toll on the Soviets. Ninety-five percent of all Soviet hepatitis patients had hepatitis A, but there are now vaccines against it. The remaining 5 percent contracted hepatitis E (Grau, 1997). Immunoglobulin has been used for over 40 years to protect against hepatitis A infection (DIA, AFMIC, 2001). "It is safe and highly effective if given before or within 14 days of exposure. The protection provided is immediate but relatively short-lived (approximately one month per ml)," so prolonged periods of deployment require repeated injections (DIA, AFMIC, 2001). Hepatitis E, formerly called non-A, non-B hepatitis, is a water-borne infection, and is found in epidemics and sporadic cases. "The disease primarily affects young adults, is clinically similar to hepatitis A, and does not lead to chronic disease" (DIA, AFMIC, 2001). There is no vaccine against hepatitis E, and immunoglobulin prepared in Europe or the United States does not give protection (Benenson, 1995). As for many other enteric infections, avoidance of contaminated food and water is the only effective protective measure.

As of March 2000, the World Health Organization (WHO) considered several provinces cholera endemic. From May through July 1999 there were 14,402 cases of severe diarrhea, including cholera cases. In July 2002, there was an outbreak of 6,691 diarrhoeal diseases, including three deaths in Kabul (UN, WHO, July 2002). The most affected areas were Kabul province and the central region. Also reported in September 2000 was an outbreak of cholera in the southern, western, and northern regions (in the provinces of Kandahar, Badghis, and Jowzjan). To date, 4,500 cases and 114 deaths have been reported (UN, WHO, July 2002).

Figure 12.2 Distribution of Poliomyelitis (Acute Paralysis)

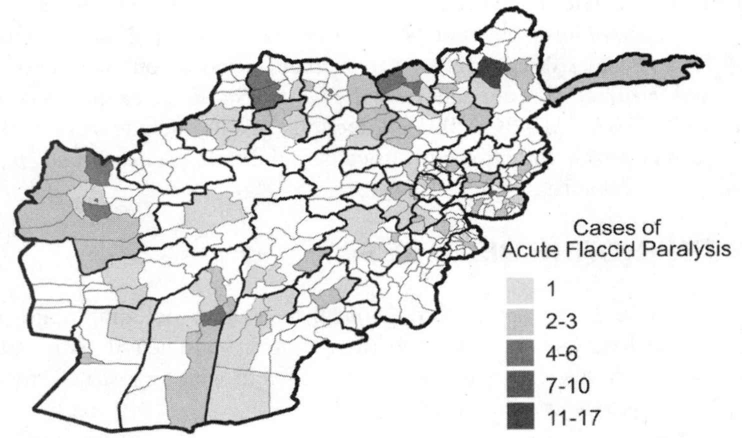

Cases of
Acute Flaccid Paralysis

1
2-3
4-6
7-10
11-17

Source: Data from United Nations, Food and Agriculture Organization, 2001

Poliomyelitis (Figure 12.2) is a viral infection occurring in areas where sanitation is poor. It is endemic in many parts of Afghanistan. Current immunizations and vaccinations can protect nongovernment workers and travelers from this disease. Outbreaks occur frequently and the WHO has made the eradication of polio in Afghanistan a high priority because the country is one of the last remaining source regions of polio (UN, WHO, 2001). Unfortunately, the lack of polio immunization can be attributed in part to the rural and isolated character of Afghanistan's settlement patterns. The disease causes paralysis, most often in the lower extremities. The larger urban areas of Afghanistan have the most concentrated cases of poliomyelitis because it is a disease that can be contracted by proximity to hosts of the virus. It should be noted, however, that the overall ratios of hosts per population is higher in rural areas owing to the decreased availability of medical care.

VECTOR-BORNE DISEASES

During the warmer months of May to November, the climate and ecological habitat support large populations of arthropod vectors, including mosquitoes, ticks, and sand flies (DIA, AFMIC, 2001). Significant disease transmission is sustained countrywide, including urban areas. Serious diseases may not be recognized or reported due to the lack of surveillance and diagnostic capability.

"Malaria is the major vector-borne risk in Afghanistan, capable of debilitating a high percentage of people for up to a week or more" (Benenson, 1995).

Figure 12.3 Distribution of Malaria Vivax

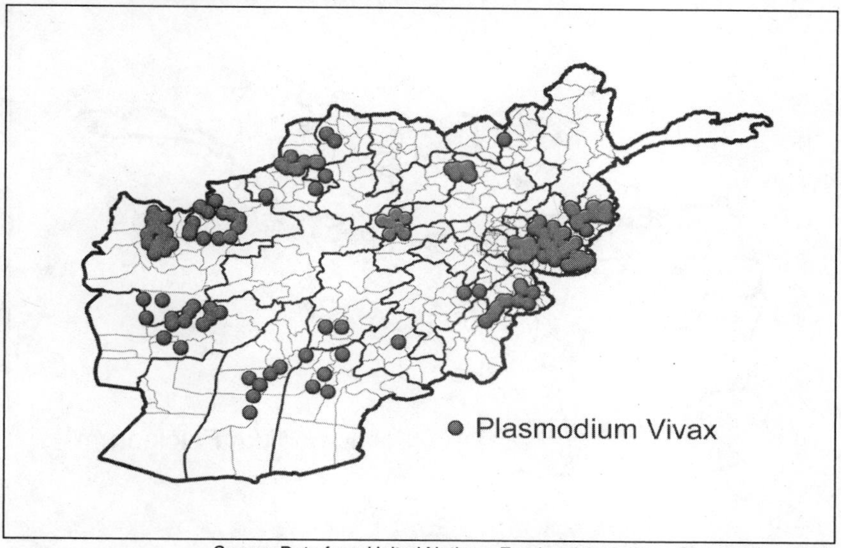

Source: Data from United Nations, Food and Agriculture Organization, 2001

Two forms of malaria exist in Afghanistan. Malaria vivax is generally not life-threatening, except in the very young, the very old, and in patients with concurrent disease or immunodeficiency (DIA, AFMIC, 2001). Malaria falciparum is the most serious form of malaria. It causes death in 10 percent of children who contract the disease and in adults who are not immune. The distribution of both forms of malaria is consistent with areas under 2,000 m (6,500 ft.) in elevation, as mosquitoes are not adapted for cool temperatures. Therefore, people who live in high altitudes would have little concern for mosquito protection. There will be concern, however, for cold weather–related health problems, addressed later in this chapter. The distributions shown in Figure 12.3 and 12.4 are explained through climate and settlement patterns in Afghanistan (United Nations, 2001). Malarial infection exists where there is sufficient water, mild to warm temperatures, and where there is human settlement. Since a large percentage of people live in the valleys of Afghanistan near sources of water, the prime locations for malarial infection include the cities of Jalalabad, Kabul, Khowst, Bamian, Gizab, Mazar-i-Sharif, Faryab, Badghis, Herat, Farah, Lashkar Gah, and Kandahar. People traveling in these regions should be supplied with prophylactic drugs such as chloroquine and mefloquine.

RESPIRATORY DISEASES

The primary respiratory diseases that jeopardize the Afghan population are meningococcal meningitis and tuberculosis. Data for meningitis are scarce, but the

Figure 12.4 Distribution of Malaria Falciparum

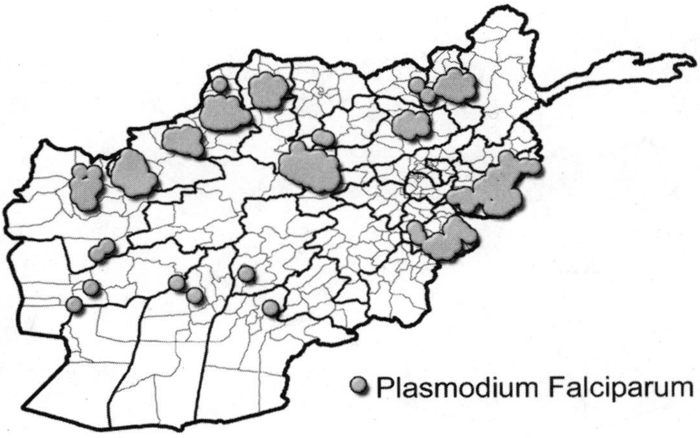

○ Plasmodium Falciparum

Source: Data from United Nations, Food and Agriculture Organization, 2001

WHO describes the risk distribution as countrywide. As of 1997, the estimated percentage of Afghans having tuberculosis was 35 percent (UN, WHO, 2001).

NUTRITION

The relationship between food and its availability to an individual for actual consumption depends on a range of intervening factors. Among these factors are family income, gender, age, season of the year, government regulations, transportation technology, and cultural factors such as dietary restrictions, taboos, and preferences. Most Afghan people are poor, and even if significant supplies of foodstuffs were available, they could not afford to buy them. Afghanistan also has a very high *dependency ratio*, further reducing the purchasing power of the working population. Afghanistan depends on a surplus of grain at harvest time to carry it through the winter months. The recent drought (Figure 12.5) crippled the government's ability to feed its population and required international intervention of food assistance and aid (USAID, 2001). When grain is received, the topography and poor transportation infrastructure make adequate distribution difficult to accomplish. Ultimately, the grain sometimes rots and is destroyed, or sometimes it is even eaten in a rotting stage, leading to more sickness. During the Taliban era cultural prejudice also came into play, and sometimes shipments did not make it past ethnic Pashtun-controlled grain distribution centers (USAID, 2001).

Medical Geography

Figure 12.5 Drought Areas

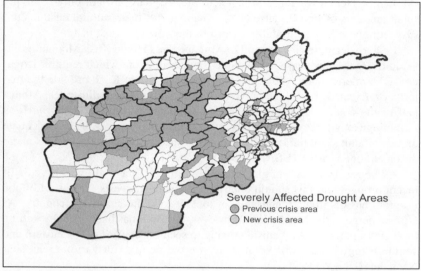

Source: United Nations, 2001

A staple food product that a large majority of the population survives on is bread made of wheat grain. A unique example of the triangle of human ecology on Afghanistan's health is liver toxicity caused by contaminated wheat. Recently, over 400 cases of liver toxicity have been reported in Herat Province (UN, WHO, 2001). During drought conditions, seeds from a toxic wild plant (charmac) compete equally with wheat seeds and both the wheat and charmac are unintentionally harvested leading to toxic contamination of wheat products. Some wheat is given to milk-producing animals, which are also infected, and the toxicity is passed on as humans consume the milk of the animal.

The impact of *malnutrition* is not easily measured, but recent studies have shown significant impacts on the people of Afghanistan. For example, iron deficiency anemia may affect over 75 percent of the population (UN, WHO, 2001). Iron deficiency in Afghanistan is another great example of the triangle of human ecology at work. Afghanistan's habitat is not conducive to large-scale beef production, and given the recent drought, there is little beef to go around. For religious reasons, Afghans do not eat pork, so this meat is not a viable source of iron, either. The dietary intake of iron is mainly through eating wheat bread. Whereas beef contains between 4 and 8 milligrams of iron per serving, Afghan wheat bread contains only 1.7 milligrams of iron per equal serving. In other words, one would have to eat an entire loaf of wheat bread to ingest the same amount of iron present in a small serving of beef. Compared to contemporary Western diets, Afghans have very low levels of iron. Also, Afghans have a cultural practice of drinking tea with meals. The caffeine in tea reduces the human body's ability to absorb iron by as much as 87 percent (McKinley Health Center, 2001). Therefore, Afghans are not only receiving small amounts of iron because of their habitat, but their cultural customs unsuspectingly reduce their ability to absorb the little they take in.

Sharp decreases in snowfall in Afghanistan's Hindu Kush Mountains has led to decreased productivity in staple food production, which requires irrigation. The intermontane valleys and rivers of the Hindu Kush provide the majority of irrigated arable land in Afghanistan, as Figure 12.5 illustrates. Almost half of the country's inhabitants are underweight, a result of *undernutrition* stemming from the severe droughts over the past two years. A lack of adequate dietary vitamins and minerals contributes to susceptibility to the various maladies mentioned earlier (USAID, 2001).

The recent rains and glacial melting in 2003 have alleviated years of recent drought and this should contribute to better harvests. The World Food Program has distributed thousands of tons of food aid across the country. As crises emerge in the Middle East, however, the WFP has had difficulty getting food aid to places in Afghanistan owing to security concerns. The eastern and southern regions are still volatile, as are the poppy cultivation areas near Nangahar (UN, WFP, 2003).

The United Nations has estimated that 3.8 million people inside Afghanistan are at risk from famine and that the food deficit is over 2 million tons of wheat (UN, FAO, 2001). This has significantly impacted human migration patterns and refugee settlement inside Afghanistan. There are an estimated 500,000 rural people who have left their homes for urban centers and an additional 200,000 people who have fled for Pakistan and Iran (UN, FAO, 2001). The country's underdeveloped urban network is incapable of absorbing large numbers of migrants from rural areas. Without adequate infrastructure and a social "safety-net," local government officials (who must plan for feeding and sheltering the migrants) are easily overwhelmed.

CIVILIAN HEALTH CARE

Cultural differences among the different ethnic groups of Afghanistan contribute to changes in its medical geography. "The civil war has disrupted health care, resulting in international and nongovernmental organizations (IOs and NGOs) supplanting the Ministry of Public Health in providing basic medical and surgical services throughout Afghanistan" (UN, FAO, 2001). The quality of emergency medical care is good but limited. Medical personnel are not trained to U.S. standards. Prior to the Taliban period, women comprised 60 to 80 percent of the medical workforce. Today the number is negligible. Under the Taliban, women were forced to work in more traditional roles at home.

Owing to fundamentalist Islamic policies of the Taliban, women could not attend medical schools as students or instructors, or fill administrative and technical positions in hospitals. The country's medical schools were incapable of graduating enough males to operate public hospitals (UN, FAO, 2001). In the foreseeable future, medical school graduates will not be as well trained as their predecessors because of the lack of qualified male instructors. The quality of the workforce will further deteriorate as older physicians and technicians retire. On a hopeful note, Afghanistan's new government may integrate women back into the medical fields to alleviate the current situation.

Medical facilities are operated exclusively by humanitarian organizations or receive some degree of personnel and/or materiel support from these organizations. There are no medical facilities in the country that offer comprehensive medical and surgical services. In Kabul and Jalalabad, sufficient numbers of public and private medical facilities exist to provide citizens with basic health care, but facility damage through looting, inadequate staffing, and lack of funds severely limits the operational capacity and capabilities of most other hospitals (UN, FAO, 2001).

Pharmaceuticals and medical supplies are imported by NGOs from Europe. The Ministry of Public Health provides only token quantities. The International Committee of the Red Cross (ICRC) transports medical materiel between combatants, only canceling convoys when fighting closes roads or airports. Although major combat operations have concluded, bandits and general lawlessness have continued to disrupt or deter medical support in some regions of the country. The ICRC maintains an 8-month supply of surgical supplies and Doctors Without Borders (*Medecins Sans Frontieres*) maintains a 6-month supply of emergency medical materiel for contingencies. The blood provided by IOs and NGOs is safe for use. Blood collected locally is not safe (UN, FAO, 2001).

CLIMATE AND TOPOGRAPHIC CONSIDERATIONS

Climate and topographic variability are two concepts that impact the health of Afghanistan's people. First, most of the country experiences extreme diurnal and annual temperatures, which are potential *insults* that influence people's health. The annual temperature range between winter and summer are lowest in the lower latitudes and highest in the high altitudes (Figures 12.6 and 12.7 [C-5]). The warm, summer temperatures of the low altitudes makes agriculture more productive, but warm temperatures also provide a natural habitat for the mosquito, producing malarial diseases. Alternatively, the cool, summer temperatures at higher altitudes shorten the growing season, limiting the available food supply for people who live in the high valleys.

Afghanistan's topography produces extreme diurnal temperature differences in the higher altitudes, which forces people to adapt to extremes. Differences between daytime and nighttime temperatures can be as much as 60 degrees Fahrenheit. Consequently, the cultural landscape of Afghanistan reflects the human adaptation to the environment—natural building materials such as straw, stone and earth help to moderate the inside temperatures of houses. A health concern related to these types of building materials is that the latter are difficult to

Medical Geography

Source: Cherie A. Thurlby

Afghani children from the village of Aroki, in the province of Kapisa, await medical screening and treatment by U.S. military medics.

keep clean; they produce dust, which induces respiratory problems, and they attract other living organisms that may carry a wide range of diseases. Unfortunately, most people cannot afford alternative construction materials such as timber, concrete, or steel. Lastly, the topography coupled with the winter season limits transportation of food and medical supplies during that time of year. Careful planning during the summer is necessary for stocking the mountainous regions with enough food and medical supplies to last until April.

SUMMARY

Medical geography plays a key role in understanding Afghanistan's culture and the interaction between people and their environment. Any medical geographic analysis requires a synthesis of information taken from nearly every subfield of geography. The health of the Afghan people can be attributed, but not limited to, the subject matter contained in virtually every chapter of this book. Afghanistan needs to make great strides in the near future if it is to overcome the nutritional deficiency and disease that plagues its people. Unfortunately, access to health care is a major problem that cannot be easily solved because of a severe shortage of medical personnel, the lack of a national budget to provide social services (including health care), continuing instability and infighting that hinders the mobility of organizations such as Doctors Without Borders, and a complete lack of infrastructure outside of Kabul, making it difficult to access the country's rural population.

13

Conclusion

Eugene J. Palka

Regional geographies are always in the process of evolving because regions constantly change through time. Regions are actually quite arbitrary constructs, since they are based on the criteria that geographers use to define them. Ideally, we discover some degree of homogeneity within the regional boundaries so that we can identify and explain the likenesses and differences between the study area and other places or regions. The political boundaries that define the territorial extent of the political entity of Afghanistan also served as the regional boundaries in our endeavor. As we have discussed previously, there is a long history of cultural, political, and economic interaction across the political boundaries of the state, particularly with the adjacent countries. Nevertheless, we have chosen to restrict the scope of our effort and focus on Afghanistan, fully aware that the country exists within the larger context of Central Asia.

We have attempted to take a current snapshot of Afghanistan and organize and synthesize the data from a range of subfields in a coherent fashion in order to develop a geographic portrait of the country. Like any other state, Afghanistan currently reflects the events and processes of its past, and ongoing activities will undoubtedly shape its future. It would be premature, however, to project the long-term effects of current physical and human activities. On one hand it is possible that some of our findings may become dated based on military operations, humanitarian assistance, and nation-building efforts that are ongoing. On the other hand, Afghanistan has a long history of successfully resisting change despite numerous interventions that have penetrated its permeable borders from the outside and the continual power struggles that have occurred internally.

Afghanistan is physically and culturally complex. The country's inaccessible location, rugged terrain, harsh climate, cultural complexity, political chaos, and poor infrastructure have impeded various types of military operations to a degree over the past year and a half, but perhaps more important, these same attributes have combined to hinder peace operations and nation-building. It will be challenging for the new government to exercise uniform control over such a rugged land without the necessary transportation networks and communications systems. Tribal allegiance and linguistic diversity further

97

Conclusion

Source: Jeremy Colvin

Two young girls of the Kofi Sofi district guide their mule away from a local water source.

complicate matters. And, it will be difficult for many to accept a new political system from a fragile government that lacks the financial resources to provide for its people.

We have employed the regional method of geography in order to better understand the distinguishing features and characteristics of the country. It may be an overly ambitious task to attempt to grasp the totality of such a diverse place in a brief publication. Nevertheless, we have sought to recognize areally associated features that give rise to a distinct regional pattern that occurs within the country. Our end product is, of course, very much of a generalization. Yet by integrating various aspects of the physical and human character of the place, we hope to differentiate it from other countries and regions of the world. If we have succeeded in providing a useful reference for students, academics, government and military personnel, then we have accomplished our main objective.

Contributors

ABOUT THE EDITOR

Eugene J. Palka received a B.S. from the U.S. Military Academy at West Point in 1978. He earned an M.A. in geography from Ohio University and a Ph.D. from the University of North Carolina at Chapel Hill. He is a professor of geography and the deputy head of the Department of Geography & Environmental Engineering at the U.S. Military Academy at West Point. He has authored or coauthored seven books, more than a dozen book chapters, several instructor's manuals to accompany college textbooks, and more than forty articles on various topics in geography.

ABOUT THE CONTRIBUTORS

Peter A. Anderson holds bachelor's and master's degrees from the State University of New York at Albany. He earned his doctorate from the University of Utah in 1994. He is currently an assistant professor of geography at the United States Military Academy at West Point.

Dennis D. Cowher received a B.S. from the United States Military Academy in 1992. He subsequently earned an M.S. in geography from Penn State. He is currently an assistant professor of geography at the United States Military Academy at West Point.

James B. Dalton received his undergraduate degree from Providence College in 1979. He holds advanced degrees from the Naval War College and Gannon University. He earned his doctorate from the University of Minnesota in 2001. He is currently an assistant professor of geography at the United States Military Academy at West Point.

Jeffrey S. W. Gloede received his B.S. from the United States Military Academy in 1992 and an M.S. from the University of Missouri-Rolla. He is currently an assistant professor of geography at the United States Military Academy at West Point.

Brandon K. Herl received a B.S. in geography at the United States Military Academy in 1990, and an M.S. from Colorado State University. He is currently an assistant professor of geography at the United States Military Academy at West Point.

Wendell C. King earned his doctorate in environmental engineering from the University of Tennessee in 1988. He also holds degrees from Tennessee Technological University and the Naval War College. A Professional Engineer, he is currently professor and head of the Department of Geography & Environmental Engineering at the United States Military Academy at West Point. He has authored numerous professional and technical publications.

Albert A. Lahood earned a B.S. from Salem State College in 1992 and an M.A. in geography from Syracuse University. He is currently an assistant professor of geography at the United States Military Academy at West Point.

Andrew W. Lohman received a B.S. in geography from the United States Military Academy in 1989 and an M.A. from the University of South Carolina in 1999. He was a member of the geography faculty at the United States Military Academy at West Point from 1999 to 2002.

Jon C. Malinowski received a B.S. in foreign service from Georgetown University and earned an M.A. in geography and a Ph.D. from the University of North Carolina at Chapel Hill. He is currently an associate professor of geography at the United States Military Academy at West Point and the coauthor of several books and numerous publications.

Patrick E. Mangin graduated from the United States Military Academy with a B.S. in 1990 and earned an M.A. in geography from the University of Minnesota. He is currently an assistant professor of geography at the United States Military Academy at West Point.

Richard P. Pannell is a 1989 graduate of the United States Military Academy and a 1999 graduate of the University of Wisconsin at Madison, where he earned a Master of Science degree in geography. He served on the geography faculty at the United States Military Academy at West Point from 1999 to 2002.

Matthew R. Sampson graduated from the United States Military Academy with a B.S. in 1991 and holds a Master of Arts degree from the University of Kansas and a Master of Education degree from Drury College. He currently is an assistant professor of geography at the United States Military Academy at West Point.

Wiley C. Thompson is a 1989 graduate of the United States Military Academy and a 1999 graduate of Oregon State University, where he earned a Master of Science degree. He served on the faculty of the United States Military Academy at West Point from 1999 to 2002.

Bibliography

Afghanistan Country Review, 2001–2002. CountryWatch.com. http://www.countrywatch.com/ (Accessed: 30 September 2001).

Afghanistan Country Review, 2003. CountryWatch.com. http://www.countrywatch.com/ (Accessed: 27 Mar 2003).

Afghanistan Political Map. Perry-Castañeda Library Map Collection, The University of Texas of Austin. http://www.lib.utexas.edu/maps/afghanistan.html (Accessed: 1 October 2001).

Afghanistan Transportation System Map. 1980. U.S. Military Academy Map Library (source information unknown).

Afghanistan. Microsoft® Encarta® Online Encyclopedia, 2001. http://encarta.msn.com (Accessed: 29 September 2001).

Ahrari, M. Ehsan. 2001. *Jihadi Groups, Nuclear Pakistan, and the New Great Game*. Carlisle: Strategic Studies Institute.

Air Force Combat Climatology Center. 1995. Operational Climatic Data Summary for Afghanistan, Jan 1995. https://www2.afccc.af.mil/cgi-bin/index_mil.pl?aafccc_info/products.html (Accessed: 1 October 2001).

_____. 2001. Narratives for Afghanistan. https://www2.afccc.af.mil/cgi-bin/index_mil.pl?aafccc_info/products.html (Accessed: 1 October 2001).

_____. 2001. Köppen Climate Classification. Geographic Information System Data (GIS). https://www2.afccc.af.mil/cgi-bin/indexmil.pl?aafccc_info/products.html (Accessed: 1 October 2001).

Allen, Nigel J.R. 2001. "Defining Place and People in Afghanistan." *Post-Soviet Geography and Economics*. 42(8): 545–560.

Amu Darya. Microsoft® Encarta® Online Encyclopedia, 2001. http://encarta.msn.com (Accessed: 29 September 2001).

Arianae. 2001. Afghanistan's Climate, Plants and Animal life. http://www.arianae.com/ctryclimate.asp (Accessed: 30 September 2001).

Awn, Peter J. 1984. Faith and Practice. *Islam: The Religious and Political Life of a World Community*. Edited by Marjorie Kelly. New York: Praeger.

Bandyopadhyay, J. 1992. "The Himalaya: Prospects for and Constraints on Sustainable Development." *The State of the World's Mountains: A Global Report*. Edited by P. B. Stone. London: Zed Books. 93–126.

Barry, Roger G. and Richard J. Chorley. 1998. *Atmosphere, Weather and Climate*, 7th Edition. London and New York: Routledge.

Beniston, M. 2000. *Environmental Change in Mountains and Uplands*. London: Arnold Publishers.

Benenson, Abram S. Ed. 1995. *Control of Communicable Diseases. Army Field Manual 8–33*. 16th Edition. Washington, D.C.: American Public Health Association.

Britannica.com Inc. 2001. Various articles. In *Britannica 2001 Standard Edition CD-ROM*. Chicago: Britannica.com Inc.

Central Intelligence Agency (CIA). 2001. *CIA World Factbook*. Washington, DC: Government Printing Office.

Bibliography

____. *The World Factbook 2001*, Washington, D.C.: The Library of Congress, 2001. http://www.odci.gov/cia/publications/factbook/index.html (Accessed: 1 October 2001).

____. *The World Factbook 2002*, Washington, D.C.: The Library of Congress, 2002. http://www.odci.gov/cia/publications/factbook/index.html (Accessed: 3 March 2003).

De Blij, H.J., and Peter O. Muller. 2001. *Geography: Realms, Regions, and Concepts*, 10th Edition. New York: John Wiley.

Defense Intelligence Agency (DIA). Armed Forces Medical Intelligence Center (AFMIC). 2001. "Medical Environmental Disease Intelligence and Countermeasures (MEDIC)." CD-ROM.

Editorial. 2001. "The Afghan Iconoclasts." *Economist* 8 March.

Edwards, M. 2001. "The Adventures of Marco Polo." *National Geographic* 199: 2–31.

Encyclopedia Britannica Online. 2001. *Afghanistan.* http://www.search.eb.com/bol/topic?eu=108466&sctn=1 (Accessed: 1 Oct 2001).

Energy Information Administration 2001. *Afghanistan.* http://www.eia.doe.gov/emeu/cabs/afghan.html (Accessed: 25 September 2001).

English, P. W. 1984. *World Regional Geography: A Question of Place.* New York: John Wiley.

Ethnologue. 2000. *Ethnologue: Volume 1 Languages of the World*, 14th Edition. Edited by Barbara F. Grimes. Dallas, TX: Summer Institute of Linguistics.

Getis, A., J. Getis, and J. D. Fellmann. 2001. *Selections from Introduction to Geography.* Boston: McGraw-Hill.

Glassner, Martin I. 1996. *Political Geography*, 2nd Edition. New York: John Wiley.

Grau, Lester W (LTC-Ret) and William A. Jorgensen (MAJ). 1997. "Beaten By the Bugs: The Soviet-Afghan War Experience." *Military Review.* Fort Leavenworth KS: Command and General Staff College, Foreign Military Studies Office. http://www-cgsc.army.mil/milrev/english/novdec97/grau.htm (Accessed: 30 September 2001).

"Hindu Kush." Microsoft® Encarta® Online Encyclopedia, 2001. http://encarta.msn.com (Accessed: 29 September 2001).

Hudson, J. and E. Espenshade, Jr., eds. 2000. *Goode's World Atlas*, 20th Edition. New York: Rand McNally.

"Impact of the Earthquake." 1998. http://www.oxfam.org.uk/atwork/emerg/afghan0698.htm (Accessed: 30 September 2001).

Jackson, B., Christina Surowiec, Julie Zhu and et al. 2001. *Country Watch: Afghanistan 2001–2002.* Houston: CountryWatch.com.

Kaplan, Robert D. 2000. "The Lawless Frontier." *Atlantic Monthly* September: 66–80.

Kiple, Kenneth F. 1993. *The Cambridge World History of Human Disease.* Cambridge: Cambridge University Press.

Khyber Pass. Microsoft® Encarta® Online Encyclopedia, 2001. http://encarta.msn.com (Accessed: 29 September 2001).

"Land and Resources." 2001. http://www.afghan-web.com/geography/lr.html (Accessed: 29 September 2001).

Lieberman, Samuel S. Afghanistan: Population and Development in the "Land of Insolence" *Population and Development Review*, Volume 6, Issue 2 (June 1980), 271–298.

Lonely Planet, 2001. http://www.lonelyplanet.com/destinations/middleeast/afghanistan (Accessed: 29 September 2001).

Margolis, Eric. 2000. *War at the Top of the World: The Struggle for Afghanistan, Kashmir, and Tibet.* New York: Routledge.

Marvel, B. 2001. "U.S. Overestimating Size of Caves, Experts Believe." *Dallas Morning News* Sec A:13, 16 October.

Matinuddin, Kamal. 1999. *The Taliban Phenomenon.* Oxford, UK: Oxford University Press.

Mayer, Ann Elizabeth. 1984. "Islamic Law." *Islam: The Religious and Political Life of a World Community.* Edited by Marjorie Kelly. New York: Praeger.

McKinley Health Center, September 2001. *Dietary Sources of Iron.* http://www.mckinley.uiuc.edu/handouts/dietiron.html (Accessed: 30 September 2001).

McKnight, Tom L. and Darrell Hess. 2002. *Physical Geography: A Landscape Appreciation*, 7th Edition. Upper Saddle River, NJ: Prentice Hall.

Meade, Melinda S., John W. Florin, and Wilbert M. Gesler. 1988. *Medical Geography.* New York: Guilford Press.

Office for the Coordination of Humanitarian Affairs (OCHA). Afghanistan-Earthquake. OCHA Situation Report February 1998. http://stone.cidi.org/disaster/98a/0033.html (Accessed: 30 September 2001).

Palka, E.J. 2001. "Introduction to Urban Geography." *Physical Geography Study Guide*, 3d Edition. West Point, NY: United States Military Academy.

_____. 2001. Medical Geography. *Physical Geography Study Guide*, 3d Edition. West Point, NY: United States Military Academy.

_____. 2001. The Physical Setting of Afghanistan. *Post-Soviet Geography and Economics*. 42 (8) 568–577.

Physical Geography Study Guide. 2001. West Point, NY: U.S. Military Academy.

Polk, William R. 1991. *The Arab World.* Cambridge, MA: Harvard University Press.

Polo, Marco (original date unknown). *The Travels of Marco Polo, the Venetian.* Translated and edited by William Marsden and re-edited by Thomas Wright. 1948. Garden City: Doubleday Press.

Rashid, Ahmed. 2001. *Taliban, Militant Islam, Oil, and Fundamentalism in Central Asia.* Yale University Press.

Rubenstein, James M. 1996. *The Cultural Landscape: An Introduction to Human Geography.* Prentice Hall, Upper Saddle River, NJ.

Saba, D. 2001. Geography: Land and Resources. http://www.afghan-web.com/geography/environment.html (Accessed: 1 October 2001).

Schindler, J. S. 2002. "Afghanistan: Geology in a Troubled Land." *Geotimes.* http://www.agiweb.org/geotimes/feb02/feature_afghan.html (Accessed: 2 April 2003).

Shah, Saira. 2001. Nightmare World. *The World Today.* Aug/Sep 2001: 25–26.

Shroder, J. T. 2001. Contribution to "Cave War." *Lehrer News Hour* (Public Broadcasting Service). http://www.pbs.org/newshour/bb/asia/afghanistan/caves_12-10.html (Accessed: 2 April 2003).

Stump, Roger, W. 2000. *Boundaries of Faith: Geographic Perspectives on Religious Fundamentalism.* Lanham, Maryland: Rowman & Littlefield.

Tarrock, Adam. 1999. The Politics of the Pipeline: the Iran and Afghanistan Conflict. *Third World Quarterly* 20(4): 801–820.

"Thousands Reported Killed in Afghanistan Quake." CNN, 6 February 1998. http://www10.cnn.com/WORLD/9802/06/afghanistan.quake (Accessed: 30 September 2001).

United Nation. 1999. Security Council Resolution 1267. http://www.un.org/Docs/scres/1999/99sc1267.htm (Accessed: 27 September 2001).

_____. 2000. Security Council Resolution 1333. http://www.un.org/Docs/scres/2000/res1333e.pdf (Accessed: 27 September 2001).

_____. 2001. *Afghanistan Online: Cooking/Food.* http://www.afghan-web.com/culture/ (Accessed: 30 September 2001).

_____. 2001. Food and Agriculture Organization (FAO). September 2001. *Assistance Afghanistan Online Mapping Program.* http:// www.pcpafg.org (Accessed: 30 September 2001).

_____. 2001. Food and Agriculture Organization (FAO). *Global Information and Early Warning System Special Report No 318, Afghanistan, 20 September 2001.* http://www.fao.org/WAICENT/faoinfo/economic/giews/english/alertes/2001/SRAFGH31.htm (Accessed: 1 October 2001).

_____. 2001. *Global Illicit Drug Trends 2001.* http://www.un.undcp.org/global_illicit_drug_trends.html (Accessed: 25 September 2001).

_____. 2001. *The State of the Afghan Economy.* http://www.afghan-web.com/economy/econstate.html (Accessed: 26 September 2001).

_____. 2001. World Health Organization. *Communicable Diseases in Afghanistan.* http://www.who.int/disasters/emergency.cfm?emergencyID=2&doctypeID=14 (Accessed: 1 April 2003).

_____. 2002. World Health Organization. *Acute Watery Diarrhoeal Syndrome in Afghanistan.* http://www.who.int/disease-outbreak-news/n2002/july/17july2002.html (Accessed: 1 April 2003).

_____. 2003. World Food Program. *Emergency Report No. 11.* http://www.wfp.org/newsroom/subsections/emergencies_report.asp?id=1 17 (Accessed 1 April 2003).

United States Agency for International Development (USAID). 2001. *Afghanistan— Complex Emergency Information Bulletin #2 (FY 2001)* http://www.usaid.gov/ (Accessed: 30 September 2001).

_____. 2002. USAID Welcomes Increase in Crop Production In Afghanistan. Washington, DC: Press Office. http://www.usaid.gov/press/releases/2002/pr020814.html (Accessed: 1 April 2003).

United States Census Bureau. 2001. http://www.census.gov/ipc/www/idbsum.html (Accessed: 30 September 2001).

United States Department of Health and Human Services. Center for Disease Control. 2001. *CDC Health Information for Travelers to the Indian Subcontinent.* http://www.cdc.gov/travel/indianrg.htm (Accessed: 30 September 2001).

Various Russian 1:200,000 Scale Topographic Line Maps. 1985. Perry-Castañeda Library Map Collection, The University of Texas of Austin. http://www.lib.utexas.edu/maps/afghanistan.html (Accessed: 1 October 2001).

Weisbrode, Kenneth. 2001. "Central Eurasia: Prize or Quicksand? Contending Views of Instability in Karabakh, Ferghana, and Afghanistan." *International Institute for Strategic Studies.* Adelphi Paper 338. New York: Oxford University Press.